OWLS

For Jamie, Sebastian and Lola

First published in the United States of America in 2012 by Cornell University Press.
Simultaneously published in the United Kingdom by Bloomsbury Publishing.

Library of Congress Cataloging-in-Publication Data
Taylor, Marianne, 1972-
Owls / Marianne Taylor.
p. cm.
Includes bibliographical references and index.
ISBN 978-0-8014-5181-2 (cloth : alk. paper)
1. Owls. I. Title.

QL696.S8T393 2012
598.9'7--dc23   2012023191

Commissioning editor: Jim Martin
Project editor: Jasmine Parker
Design by Nicola Liddiard, Nimbus Design

Printed in China

Cloth printing   10 9 8 7 6 5 4 3 2 1

OWLS

Contents

Above: Great Grey Owl
Left: Boreal Owl

Introduction

Birds are a highly visible part of the natural world. They grab our attention with their bright colours, showy flight, melodious songs, and bold and cheeky characters. However, one of the most well-known and loved group of birds possesses none of these traits. They are the owls, secretive nocturnal birds of the forest, patterned in cryptic tones and borne on soundless wings.

So why have owls captured our imaginations so thoroughly? The first and most obvious reason is staring you in the face when you go eye to eye with an owl – they look like us. Eye contact is crucial for us to relate to each other, and we naturally feel more drawn to animals that can return our gaze with two forward-pointing eyes. However, among birds most species have side-mounted eyes for a wider field of view. Only owls can truly look at us as we look at them.

HIDDEN LIVES

For most of us, though, moments like this may only happen once or twice in our lives. Owls' lives do not intersect very often with those of humans – at least not with humans who live in towns and cities. Even keen bird-watchers who spend their days roaming the country-side, seeing hundreds of birds, treasure every owl encounter for the special moment that it is. The rest of us are most likely to see an owl at dusk, a fluttering shadow dodging the headlight beams or a silhouetted lump on a tree branch. More likely, we don't see them at all but hear them proclaiming their territories after dark, their far-carrying cries sometimes haunting, sometimes downright alarming.

So skilled are owls at pursuing unobtrusive lifestyles that new species are still being discovered today. The Serendib Scops Owl of Sri Lanka lives in rainforest that has been explored and studied by biologists for decades, but the owl's presence was only detected in 1995, when a scientist homed in on its frog-like call. Even then, it took six more years before the owl was actually seen.

Little wonder, then, that owls have inspired so much folklore and legend, and feature so prominently in fairy tales and fantasy, where they may be portrayed as bearers of wisdom or sinister harbingers of bad luck. In the real-life dramas of wildlife documentaries, their role is that of beautiful, silent assassin, targeting and dispatching their prey with clinical efficiency.

UNDERSTANDING OWLS

There are some 220 species of owls throughout the world. The majority are nocturnal, woodland birds, though there are exceptions to this. Among them are two of the most widespread bird species on Earth, the Barn Owl and the Short-eared Owl. However, many others have tiny populations and are considered to be critically endangered – in imminent danger of total extinction. Without immediate conservation efforts, these birds will soon be lost to us for ever.

Studies of some of the more common species have given us remarkable insights into the anatomical and behavioural attributes that make them so beautifully adapted to their way of life. Gradually, scientific investigation is uncovering the owl's secrets. However, in some parts of the world owls are seen more as threatening representations of dark spirits or as a source of tasty meat – both attitudes spell obvious disaster for owl populations and potentially the loss of priceless knowledge.

Studying owls can take many forms. Dissecting their pellets (indigestible food remains coughed up after feeding) reveals the content of their diet. Erecting nestboxes makes it easier to access the owlets and mark them with leg rings, allowing individual birds to be reidentified in the future – nestboxes can also significantly boost owl populations. Even captive owls, perhaps those injured in accidents that never recover enough to return to the wild, have an educational role to play – they allow adults and children to enjoy close-up study that would be impossible in the wild.

ABOUT THIS BOOK

This book aims to both celebrate and demystify these seldom-seen and misunderstood birds. It is divided into two main sections. The first section looks at owls in general, with chapters covering broad owl attributes, senses, hunting behaviour and techniques, habitat, breeding biology, conservation issues and finally the long and sometimes troubled relationship between humans and owls. The second section presents detailed individual accounts for owl species found in North America and Eurasia, including information on how to see them.

These wonderful birds share our world so discreetly that you could easily live out several lifetimes without ever seeing one. However, with a little insight, patience and luck, you could step into their realm and enjoy some unforgettable close encounters, whether abroad or at home. The chances are high that there are wild owls of one kind or another living not far from your own front door, and your observations could make a difference to their survival.

The first nine chapters of this book look at owls in general – their diversity, biology, way of life and interactions with people.

The second half of this book explores particular owl species in detail.

What makes an owl?

With a handful of exceptions, the various owl species are immediately recognisable as owls. Yet the traits that make them what they are go beyond what we can see at a glance. The owls are a very distinct group, with no very close allies in evolutionary terms, and the relative lack of diversity within this group just shows how successful the owl 'prototype' is as a recipe for survival. In this chapter we look at how owls evolved and diversified to establish themselves throughout the world.

CLASSIFICATION

It is natural for us (and perhaps other animals too) to attempt to categorise the other living things that share our environment. For human cultures living off the land, it is vital to know which plants are edible and which poisonous, which woods burn hottest, which animals are easy to catch and which are dangerous". While for developed cultures today this need may be less pressing, the interest is as strong as ever, and we are developing new techniques all the time to refine our systems of classification.

In its early days, the science of biological classification involved grouping living things according to overt physical similarities. The Swedish scientist Carl Linnaeus, working in the early 18th century, devised the first formal classification of life on Earth. His *Systema Naturae* categorised 4,400 species of animals, which were divided on the first level into mammals, birds, reptiles, amphibians, fishes, insects and 'worms' (all other invertebrates). Each of these main groups was then further subdivided. He placed the owls in a subcategory together with the raptors and the parrots – understandable enough, as all three are generally medium-sized woodland birds with prominently hooked bills.

In 1859, Charles Darwin's *On the Origin of Species* was published, showing what Linnaeus had already inferred – that all life on Earth is related, descended from a single common ancestor and, over the course of time, diversifying and branching out into a host of new forms. This new understanding gave more structure to the evolving study of classifying life on Earth, which became the science of taxonomy. Now, scientists were looking for evolutionary relationships, and also for the red herrings of convergent evolution, where unrelated animals evolve with similar general physical traits as adaptations to similar lifestyles. For example, Linnaeus initially placed whales with fishes, based on their obvious outward similarity, but later reclassified them correctly as mammals, as they breathe air and have mammary glands.

Since *On the Origin of Species,* we have a huge array of techniques to study relationships between species. We no longer need to rely on outward physical attributes to group organisms but can draw on evidence from the fossil record, distribution, hybridisation, deep anatomy, vocalisations and, most recently, molecular DNA evidence to infer the fine details of exactly when two related species 'split' from their most recent common ancestor. However, there is still much to learn, and every week some old assumption is overturned by new evidence.

With superlative hearing, excellent low-light vision and a soundless flight, the Eastern Screech Owl is a perfect night hunter.

Left: We think of owls as night birds, but many species, like the Great Grey Owl, are more active at dusk and dawn than midnight.

Taxonomy explained

Modern taxonomy divides living things into categories or taxa (singular taxon) which themselves are divided into subcategories, forming what's known as a 'nested hierarchy'. The basic unit of biology, the species, denotes a population of organisms which interbreed. Each species has a two-word or binomial scientific name – for example, the Short-eared Owl is *Asio flammeus*.

The next level up from species is the genus (plural genera), which contains a number of closely related species. All species within a genus have the same 'first name' in their scientific names, and are called 'congeners'. Congeners of the Short-eared Owl include the Long-eared Owl *Asio otus*, Striped Owl *A. clamator* and Marsh Owl *A. capensis*. In total, there are seven different species of owls in the genus *Asio*.

Moving up another level, we come to the family, containing one or several closely related genera. The genus *Asio* is classed in the family Strigidae, along with about 22 other genera, such as *Glaucidium* (the pygmy owls), *Bubo* (the eagle owls) and *Micrathene* (a genus with just one species in it, the Elf Owl *Micrathene whitneyi*). The next level up is the order, containing one or multiple closely related families. The owl order is Strigiformes, and contains just two families – Strigidae (as described above), which contains the vast majority of owl species, and Tytonidae – the barn owls, with somewhere between 15 and 19 species. Strigiformes is united with all other bird orders in the class Aves or Birds.

Every member of a species, a genus, a family and so on shares a common ancestor somewhere in the past, and every taxon represents a complete lineage, however much it may have split and diversified along the way. These groupings are called monophyletic clades, monophyletic meaning that they comprise all the descendent taxa of a single common ancestor.

Nothing in taxonomy is set in stone, meaning scientific names sometimes change. Taxonomists are constantly challenging the correct classification of not just individual species but entire classes of organisms, as new evidence about their relationships is revealed. To a certain extent, we are on a hiding to nothing in our attempts to classify the natural world, because by its very nature it is 'messy', with fuzzy rather than clear-cut boundaries between species and also at higher levels. Nevertheless, having a general idea of the system does help to understand both the evolutionary process and the nature of individual species.

The Short-eared Owl is one of the world's most widespread birds, and shares its genus with six other species.

In the case of owls, their taxonomic position is still unclear. They do form a single cohesive group, all evolved from one early owl-like lineage, but there is dispute over their relationships with the other birds. Evidence from comparing physiology and behaviour indicates that the falcons (but not other raptors) are owls' closest cousins. However, a 1990 study produced DNA evidence that aligns the owls with the nightjars, while a 2006 DNA study found that, contrary to expectations, the owls are most closely allied to the non-falcon raptors. Two different study methods – mitochondrial DNA sequencing and DNA–DNA hybridisation respectively – produced these latter two results, so there is clearly more to study.

EARLY OWLS

Fossil owls have been found in deposits that date back to the Paleocene epoch, some 60 million years ago, and only a short time (geologically speaking) after the extinction of the dinosaurs around 65 million years ago. It is therefore possible that the last dinosaurs and the first owls coexisted. Some of these primitive owls were quite unlike the owls we know today. The family Sophiornithidae, for example, were ground-birds, though still hunters, chasing down their prey on foot. Others may have been more owl-like in appearance but had yet to develop some of the specialised adaptations of modern owls. For example, modern owls usually perch with two toes pointing forward and two backwards (unlike most birds which have three toes forwards), but members of the fossil genus *Berruornis* were only just starting to develop this trait.

Going forward to a mere 25 million years ago, various species of barn owls (family Tytonidae) greatly outnumbered the 'typical owls' (family Strigidae) in Eurasia at least, but the typical owls began to spread and diversify at the same time that their main prey animals, the rodents, really began to prosper. The typical owls had become the dominant group by modern times.

'Eared' owls like this Eurasian Eagle Owl all belong to the major subfamily Striginae.

Right: Specialised for hunting in and around rivers, Blakiston's Fish Owl has unusual and imposing looks.

MODERN OWLS

Compared to some other bird orders, the owl order Strigiformes is rather homogeneous, both in outward appearance and in general behaviour. However, from species to species there are some interesting distinctions, some easily explained and others much more mysterious.

Appearance The greatest variety among owls is perhaps in size. The world's smallest owl, the Elf Owl, weighs just 40g and is no larger than a sparrow. At the other extreme are the larger species in the genus *Bubo*, with Blakiston's Fish Owl *Bubo blakistoni* and Eurasian *B. bubo* battling for the top spot. Both species can weigh up to 4.5kg, and measure up to 75cm from bill to tail-tip.

Most owls are patterned in a variety of brown and grey tones to blend in with their woodland habitat. Their colours and patterns are beautifully intricate at close range but function as perfect camouflage when the owl is roosting among the trees. Owls that live primarily in deciduous forest tend to have more brown tones to their plumage, while owls of pine woodland are greyer, because of the different bark colours of the trees. Owls of the rainforests often exhibit darker plumage. And this variety in tone can even be seen within individual species – Tawny Owls *Strix aluco* in Britain, where oaks and beeches dominate the woodland, are much browner than their counterparts living in the coniferous taiga forests of Finland.

The most striking exception to the usual colour scheme is seen in the Snowy Owl *Bubo scandiacus*, which is white to match its often snow-covered Arctic habitat. This white plumage makes the Snowy Owl very different outwardly to its close relatives the eagle owls, but 'under the feathers' it reveals its true relationships, in behaviour, voice and anatomy. Owls of more open ground, such as the Barn Owl *Tyto alba*, generally have lighter plumage tones than forest-dwelling birds.

Owl eye colour is another trait that varies across the group. Some species have black eyes, others orange or bright yellow. The reasons for the variety are unclear. The three different colours occur across and within the various owl genera – for example, the Short-eared, Long-eared and Striped Owls of the genus *Asio* are yellow-, orange- and dark brown-eyed respectively (although Long-eared Owls in North America are yellow-eyed). It has often been suggested that owl eye colour indicates the bird's activity pattern (whether it is active by day or night), but while most black and orange-eyed owls are strictly nocturnal, yellow-eyed owls may be nocturnal or diurnal.

Some species of owls have ear-tufts – feathery projections on the top corners of the head. Again, there is little rhyme or reason behind which owls have tufts and which do not. The genera *Asio*, *Otus*, *Megascops* and *Bubo*, which are all members of the subfamily Striginae, are composed partly or entirely of tufted species. See chapter 2 for various theories that attempt to explain the function of these tufts. Some owls that are not members of Striginae have distinct bumps on the outer corners of their crowns in the same position as the eared owls'

URAL OWL

Unique attributes

What are the features that tell us we are looking at an owl and not some other kind of bird? The most obvious feature to our eyes is the flat face, with forward-pointing eyes and a circular disk or 'ruff' enclosing the face. However, one group of raptors, the harriers, has a similar set of features. Harriers' faces are not as flat as those of owls, and their facial disks are less obvious, but the similarity is quite evident, and is an example of convergent evolution. Like owls, the harriers are low-level hunters and rely on sound as well as sight to locate prey – the facial disk helps to channel sound and enhance the birds' hearing.

What else? Owls have soft and well-camouflaged plumage, but so do many forest birds, most notably the nightjars, which are also nocturnal and almost soundless in flight. The two toes forward and two backward owl foot is not unique either – parrots, cuckoos and woodpeckers perch in the same way. However, combine this with the facial disk and soft cryptic plumage, and add in a few more traits such as the specialised eyes and ears, the sound-insulating microstructure of the flight feathers, and the ability to rotate one toe from its usual backwards-facing position to face forwards, and you have a suite of features that is uniquely owl.

Evolution and adaptation

How does an owl that lives in the cold end up with fully feathered legs, while an owl that hunts fish has bare legs? It is easy to look at the world today and conclude that every species was purpose-built to function successfully in its environment, but intuitively it can be difficult to understand that this purpose-building is a 'blind' process.

DNA, the genetic coding molecule present in every cell that tells an organism's body how to grow, is replicated each time a cell divides, but not always perfectly. Errors in replication are called mutations, and if a mutation happens in a gamete or sex cell, its effect is manifested in the organism that develops from that gamete. This is very common – all of us humans, for example, have 100 or more mutations that our parents did not have, many of which have no obvious effects. Mutations are random, but the process that removes 'bad' mutations is far from random: it is, quite simply, death. By the same token, 'good' mutations help an animal to survive and breed, and so are more likely to spread through a population.

So evolution works on these two factors – variability introduced by genetic mutation and natural selection that preserves only those variants that have survivability. If an organism's environment stays the same, it will become ever more adapted to do well in that environment, with small improvements being retained in its population. If its environment changes, though, it may be pushed in a new direction. If the habitat becomes colder, better-insulated individuals will survive. If it gets more arid, individuals that can manage with less water will do better. These selective pressures work to produce adaptation to environments. However, if an environmental change is very quick or very dramatic, there is a high risk of extinction as evolutionary adaptation is a slow process.

ABOVE: GREAT GREY OWL; BELOW: BROWN FISH OWL

tufts. Owls also may or may not have feathered tarsi (the exposed part of the leg) and toes – full feathering is the norm among species that live in cold climates and usually absent in the fish owls, which hunt aquatic prey.

Anatomy and physiology Owls possess a number of unusual anatomical traits. Many of these are related to their particularly specialised vision and hearing, and these are discussed in more detail in chapter 2.

Living owls often look rather shapeless and lump-like when at rest, especially when relaxed, their dense plumage giving them a very softened and rounded outline. The owl skeleton shows several features that are not obvious in an intact bird. The size and forward projection of the bill are striking, looking more reminiscent of the bill of a hawk than the 'flattened' look we associate with owls. This flattened appearance is an illusion caused by the thick feathering on the owl's face, especially around the base of the bill. The nostril openings are on the sides of the bill, except in the true hawk owls (subfamily Ninoxinae) which have them at the front.

Owls can look almost neckless, but the skeleton reveals quite a long neck with 14 small cervical vertebrae, giving the famed neck flexibility. The legs are also longer than you might expect – again, these are usually

hidden among thick plumage but their great length and reach are evident when you see an owl making a strike at prey. The wing bones are also relatively long, although this varies between species. Owls that hunt on the wing are particularly long-winged. Having large wings relative to body size and weight is referred to as 'low wing-loading' and enables the owl to remain airborne while flying very slowly, an essential trait for a bird that mainly hunts by scanning the ground with both eyes and ears.

Overall, the owl skeleton is very small relative to the apparent size of the bird, and also very light. As with other birds, some of the long bones are hollow, supported internally by a honeycomb structure. Additionally, some bones are fused together, giving extra strength. The range of skeletal adaptations for lightness without compromising too much on strength allows the owl to fly.

The owl digestive tract, unlike that of most other birds, does not contain a storage pouch or crop in the throat. Food passes straight to the first part of the stomach where digestion begins, and then on to the second part, the gizzard, where vigorous processing separates the items that can be digested from the material that cannot. The indigestible bits are regurgitated in the form of a compacted pellet. See chapter 3 for more on pellets.

Like other birds, owls have a structure incorporated

into their breathing apparatus called a syrinx, which enables them to make sounds. It is the equivalent of the larynx in mammals, and is responsible for the wonderful array of birdsong that we can enjoy, but the owl syrinx is relatively simple compared to that of songbirds. Muscles in the airways control the pitch and loudness of the sound as air is pushed out – when you watch a singing or calling owl, there is often visible swelling of the throat as the notes are forced out. Some owls use this to add a visual display to the song – as the throat swells, this puffs out the feathers there and enlarges or reveals a patch of contrastingly coloured plumage, like the white 'collar' of the Great Horned Owl *Bubo virginianus*.

Behaviour All owls are predators – making their living by catching, killing and eating other animals. The prey that they take varies according to size, with the smallest owls subsisting mainly on insects, and the largest species tackling sizeable birds and mammals. There are also owls that specialise in particular prey types, and have physical and behavioural adaptations to match. The fish owls will wait or walk alongside water and jump into the shallows to catch fish or frogs. The wood owls of the genus *Strix* hunt mainly by sound and can pinpoint the rustle of a mouse on the forest floor in total darkness or can home in on a vole moving under a

thick blanket of snow. Owls of open country often hunt on the wing and use a very distinctive searching or quartering flight just a metre or two above the ground.

Migratory behaviour is another aspect that varies between species, and also within species. Owls can survive more severe cold weather than many birds, but species that breed near the poles may be forced to move to more temperate climates in winter. Then there is the phenomenon of irruptive movements, when owls may move unpredictably from area to area in response to changes in their food supply. The same species may show migratory or irruptive behaviour in different parts of its range. Many other owl species are noticeably sedentary, occupying the same small 'home range' for their entire lives. This gives them the distinct advantage of thorough familiarity with their patch – a great aid to efficient hunting – and a regular nest-site which can be used year after year.

Owl breeding behaviour shows rather little variety from species to species, but in the Arctic species especially, availability of prey can have a very significant effect on breeding success. Populations of small rodents tend to be cyclical, with numbers peaking every few years, and owl populations that depend on these rodents follow a similar cyclical pattern. The Snowy Owl, dependent on lemmings to feed its brood, may

The Northern Hawk Owl has the speed and agility to catch birds in flight.

produce a double-figure clutch of eggs in a lemming 'boom' year, but may not breed at all if it is a very poor year. In general, the more northerly the species, the greater its reproductive potential, and small species usually produce bigger broods than the largest species. Most owls are strictly monogamous, with two parents assisting to rear one brood, but in times of plenty one male may be able to provision two nests, while in hard times one female may have two male partners. The male and female roles are fairly strict and consistent between owl species.

Few owls show much social behaviour. Outside of necessary duties associated with breeding, most owls are solitary, and certainly like to hunt alone – though several may 'work' a very productive area outside of the breeding season, they do not do so cooperatively. However, there are times when teaming up with others of the same species is a good thing. Owls that lead sedentary lives often form lasting pair bonds, and the two partners work together to defend their territory from intruders all year round. They may not spend all or even much time together, but bonding behaviours such as mutual preening may be seen between the pair at any time, not only as part of courtship. Owl-keepers also report developing affectionate bonds with their birds, indicating that the owls do enjoy companionship of one kind or another.

Some species form communal roosts in winter, benefiting from shared body heat and multiple eyes and ears to detect approaching predators. They can also watch each other's flight paths into and out of the roost, and thus gain information on where to look for prey. And, in the breeding season, some species may form loose colonies when there is a surfeit of prey. But owls are universally disliked by other birds and are subjected to angry mobbing when discovered at their roosts. There is little love lost between different owl species, either, with large species frequently attacking and killing smaller ones, perhaps to eliminate competition for prey and nest-sites.

Owls that inhabit forest have dark plumage. These Brown Wood Owls live in dense, tropical jungles.

Far left: Many animals that live in Arctic regions are white, but apart from this the Snowy Owl is anatomically very like the other owls in its genus.

Owl senses and intelligence

Paying attention – the Brown Hawk Owl's enormous eyes are obvious, but it is also equipped with impressive (though hidden) ears.

The unique shape of the owl face might be very appealing to us humans. However, evolution has shaped the owl with a wide flat face and large forward-facing eyes not to look adorable but to be an ever more efficient hunter in its particular niche. These adaptations are all related to the bird's remarkable sensory equipment, in particular, vision and hearing. Day-flying birds of prey are renowned for their acute eyesight, but for owls, especially those that hunt by night, exquisitely sensitive hearing is often even more vital.

SOUNDING OUT

No matter how good your eyes are, you still need some level of light for them to work well. The nocturnal owls have exceptionally acute hearing, and when hunting in near darkness, in the cover of a forest on a moonless night for example, they can still accurately locate their prey, by using sound. Even if the prey is moving around out of sight in long grass or dense leaf-litter, the tiny sounds of its movements are enough for the owl to get a 'fix', and it will simply punch through the cover to grab its victim. Some of the northern owls also use sound to pinpoint prey moving underneath a thick blanket of snow.

The owl ear has several adaptations that enhance its sensitivity to sound – in particular the kinds of sounds that prey animals make as they move around. Barn Owl ears are especially sensitive to sounds of 6–9kHz – the shuffles and squeaks of an active vole or mouse.

Anatomy of the owl ear

Unlike mammals, birds typically have little in the way of external ear structures – all that can be seen of the ear from the outside is a simple hole in the skin, behind the eye. Some owls, however, have a fleshy external ear flap or operculum, either in front of or behind the ear opening, or both. These are moveable, allowing the owl to channel sound into the ear – their movement can also change the shape of the owl's face.

The ear canal or auditory meatus leads to the eardrum or tympanic membrane. Sound waves cause vibration in the eardrum, which is passed on to the columella, a small bone in the middle ear. Mammals possess three such bones but in birds there is only one, and it connects the eardrum to the inner ear. The inner ear contains the cochlea, the structure responsible for receiving auditory information and passing it on to the brain. The middle and inner ear also contain the vestibular systems – structures to maintain balance and bodily equilibrium based on the position of the head relative to the body.

Left: The Great Grey Owl's expressive face seems full of wisdom, but are these birds really as thoughtful as they look?

When is an ear not an ear?

Many owls have pointed feathery tufts on the tops of their heads. A range of species from the subfamily Surniinae, from the tiny scops owls to the huge eagle owls, sport 'ear-tufts' of various shapes and sizes. Unlike the fleshy external ears of mammals, these tufts are just feathers, and have no role in directing sound to the bird's actual ear openings, which are set much lower down on the head-sides. The tufts can be raised or lowered.

So what function, if any, do they have? One theory is that of 'predator mimicry' – the ear-tufts give the owl's face a remarkably cat-like impression, which could put off potential attackers. Owls at their nests are vulnerable to predatory mammals, so looking like a fierce cat could be a real advantage. However, many tufted owls live in places where there are no predatory mammals, and some, such as the Eurasian Eagle Owl, are extremely powerful predators in their own right with no need to imitate any other. Also, when an owl is disturbed at its roost it tends to adopt a very static, upright and elongated posture, rather than any form of overt threat display.

A less exciting but perhaps more plausible explanation is that the ear-tufts provide a form of disruptive camouflage, giving the owl a head shape that is angular rather than smoothly rounded.

This angular shape is more likely to be overlooked among the mass of branches and bark patterns that forms the typical forest owl daytime roost. Being well hidden at the roost means less mobbing and harassment from small birds as well as protection from more serious threats – and indeed most of the prominently tufted owls are nocturnal, woodland-dwelling species – their open-country relatives have much smaller tufts. Additionally, some tuftless owls assume a more angular head shape when startled at their roost, by raising certain feathers.

The position of the tufts appears to be associated with the bird's mood. In alarm or anxiety, most owls have a 'concealment posture' whereby they stand slim and tall with tufts raised – this would also enhance the camouflage effect. In relaxed mood they are often flattened, but in 'attack-ready' mode when an owl is about to seize prey, they are raised. It can be inferred that tuft position could have a function in communication between members of the same species.

Interestingly, some owls also have 'false eyes', in the form of dark and light circular markings on the back of the neck which resemble a pair of eyes. The function of these markings is apparently to convince any predator approaching an owl from behind that it is being watched.

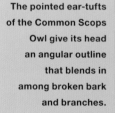

The pointed ear-tufts of the Common Scops Owl give its head an angular outline that blends in among broken bark and branches.

In some species, the openings of the ears are set asymmetrically in the head and channel sound in different directions. This means that when a sound is coming from directly above or below the owl, it reaches one ear fractionally before it arrives at the other, enabling the owl to determine exactly where the source of the sound is on the vertical plane as well as the horizontal. In the case of a few species, such as *Boreal Aegolius* funereus and Great Grey Owls *Strix nebulosa*, this ear asymmetry actually involves the skull structure as well as the location of the ear opening, and is very obvious when you examine a skull – from all angles it looks lopsided. In some owl groups, one ear opening is also clearly larger than the other.

The medulla or 'hearing centre' of the owl brain is also particularly complex, with more processing power than in similar-sized birds. The Barn Owl medulla contains around 95,000 neurons, compared to around 27,000 in a crow, meaning that the owl can make near instantaneous adjustments to its attack path as it draws in new sound information about the prey's movements. Owls also have remarkable powers of recall for sounds – studies have shown that even if distracted mid-attack, they can return later and still pinpoint the exact spot where prey was heard before.

For us mammals, the shape of our external ears helps channel sound where it's needed. In owls, the facial disk with its 'ruff' of stiffened feathers serves a similar purpose – the ear openings are situated just in front of the ruff. The hooked owl bill, projecting much less than that of a hawk or eagle, is almost flush with the bird's flat face and so does not interfere with this sound-channelling.

THE EYES HAVE IT

One of the reasons that owls are so appealing to us is that we can look into their eyes. Forward-facing eyes are unusual in the bird world. Most birds have eyes on the sides of their heads, giving a wide field of view with little or no overlap between what the two eyes can see. For animals that are more likely to be hunted than hunter, a wide field of view is vitally important, as they need to be able to detect and react immediately to potential danger from any direction.

The forward-facing eyes of owls and some other predators, with much overlap between the two eyes' fields of view, give superior depth perception – the ability to accurately judge exactly how far away an object is. This is achieved by the brain instinctively comparing the 'picture' of the object received from one eye with the slightly different version from the other. The extent to which the two images converge is determined by its distance from the eyes. Owls do have predators of their own to worry about, but they compensate somewhat for their narrow field of view by their long (albeit hidden in thick layers of feathers) flexible necks,

The Boreal Owl lacks true ear-tufts but can erect feathery bumps above its eyes.

Bird's-eye view

The anatomy of a bird's eye is not very different to that of our own. The eyeball is protected by layers of tough white sclera, which may sometimes be visible as the 'white' of the eye. At the front of the eye, the protective membrane becomes transparent, forming the cornea, which allows light through while protecting the iris. The pocket in between the cornea and iris (the anterior chamber) is filled with a clear fluid called aqueous humour.

The iris is a circular or sphincter muscle and is the coloured part of the eye. Its degree of contraction controls the amount of light that passes through the central hole – the pupil. Immediately behind the pupil is the lens, a round, transparent structure which changes shape to provide clear focus of the image received through the pupil. Light passing through the lens then crosses the interior of the main body of the eyeball (the posterior chamber) and reaches the retina on the eye's back wall. This is a layer of light-sensitive cells, containing chemicals which react when they absorb light. This chemical signal is converted into a nerve impulse – the nerve cells connect up to the optic nerve, which exits through the back of the eye and heads to the visual centre of the brain. Behind the retina, a layer called the tapetum lucidum reflects back the light, giving the retina a 'second chance' to absorb it.

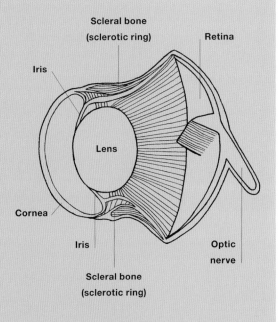

Myth-busting

It is sometimes said that owls cannot see well in normal daylight, finding it too dazzling. This is untrue – like other vertebrates they can alter the size of their pupils to control the amount of light that enters the eyes, and so avoid overloading their retinas with light. Even strictly nocturnal species can fly and manoeuvre perfectly well in daylight if they have to. However, the relative lack of colour-sensitive cone cells in their retinas means the daylight experience is probably rather monochrome.

It was once thought that infrared wavelengths, which are longer than those of light that is visible to human eyes, must be visible to owls, to account for their ability to find and catch prey in such dark conditions. However, in a study that involved shining increasingly intense infrared beams into the eyes of Long-eared Owls, the birds showed no pupil contraction, indicating that the infrared cannot be 'seen'.

Another myth surrounding owl senses is the idea that they use echolocation to navigate and find prey. Bats and some other nocturnal animals do this, making constant sounds and listening for how quickly the echoes can be heard, to build a 'sound map' of stationary and moving objects around them as they fly. However, this is not a trick that has evolved in owls. As we have seen, they can detect moving prey by sound alone, but avoiding non-moving obstacles comes down to sight and to familiarity with their territory.

BARN OWL

containing 14 cervical vertebrae (compared to just seven in mammals). This enables them to quickly turn their heads up to 270 degrees. Another adaptation is the presence of eye-like markings on the back of the head in some species, so a predator approaching from behind may be discouraged by the sense that it is actually being watched.

Owl eyes are very big relative to the head, providing a very large retina to gather as much light as possible. To accommodate such a large eye in a smallish skull, the eyeballs have developed a tubular shape rather than spherical like ours, and because they protrude some distance out of the skull (an effect that is masked by the facial feathers) the sides are protected by bony structures called sclerotic rings. This means that the eyes are rather immobile, another reason why owls need their impressive head-turning skills. They also have the very charming habit of rapidly bobbing their heads when looking at something that interests them, which helps them to refine their sense of perspective by giving both eyes a fuller picture of the interesting object and its surroundings.

SEEING IN THE DARK

Nocturnal owls have remarkable night vision. Studies on Long-eared, Barred *Strix varia* and Barn Owls have

revealed that these species can see motionless prey from a distance of 2m or more at light levels of 0.0000000678 lux. When you consider that a fully moonlit night is about 0.27 lux and an overcast night is 0.0004 lux, the owls can still see (at quite close range at least) in darker conditions than are ever likely to be encountered out in the open – a valuable back-up to their superlative hearing.

Their excellent low-light vision is partly down to having a high concentration of rod cells in the retinas. Humans have a mixture of rod and cone retinal cells, the former detecting contrast between light and dark, and the latter sensitive to colour. There are three kinds of cone cell, each responding to a different colour wavelength, together allowing us to perceive many different shades. The human retina also has a fovea – a region with lots of cone cells. When we look directly at something, its image falls on the fovea so we can see maximum colour detail. Some diurnal birds go a step further and have two foveas, plus a fourth kind of cone cell, giving superior colour discrimination. In nocturnal owls, however, there are very few cone cells, and only a rudimentary fovea. Instead, the retina is packed with light-sensitive rod cells – up to 56,000 of them per square millimetre in the Tawny Owl. Overall, a Tawny Owl's rod to cone ratio is about 1,000 to 1, compared to 20 to 1 in a human.

The light-gathering power of the retina is further enhanced by the tapetum lucidum, a film of iridescent reflective tissue behind the retina, which reflects light back towards the retina, brightening the perceived image even more. This structure, present in many nocturnal animals, is responsible for 'eyeshine' – when an animal's eyes appear to 'glow in the dark' when caught in a beam of light. Eyeshine colour can help identify an animal, owl eyeshine is usually red.

Owls that are active by day have retinal cell counts more like those of other diurnal birds, though their eyes are still proportionately very large – a Snowy Owl's eyes are as heavy as a human's. With such large eyes set in a flat face, there is more risk of injury to the eyes than with other birds with more projecting bills. Therefore,

owls will often close their eyes completely when in potentially risky situations – when feeding, passing food to a mate or a chick, or pouncing on prey. Their eyeball shape makes them long-sighted anyway, so they cannot focus on close objects.

Like other birds, owls also have a third eyelid or nictitating membrane, which is semi-transparent and is drawn across the eye from the inner side. A normal owl wink or blink, used to remoisten the eye, is performed with the nictitating membrane while the 'real' eyelids remain open. The nictitating membrane also offers protection, for example from drying out when the owl is flying fast into the wind. Unusually among birds, the upper of the true eyelids is larger than the lower, helping to block strong light.

A Ural Owl hunting by night makes the most of its low-light vision, though hearing is of more importance.

Ultra special

The ultraviolet (UV) part of the spectrum is beyond the reach of human vision but is visible to many animals, including various insects and the majority of birds. UV reflection plays an important role in mate choice for many birds, while the European Kestrel, a small falcon, is known to be able to detect voles using their UV-reflective urine scent marks. As it hovers the kestrel can see which vole runs are in use, and home in on the vole doing the marking.

It seems feasible that owls, many of which hunt similar prey, could also use ultraviolet cues in this way. However, studies have given mixed results. Most owls seem unable to see UV or near-UV light. Research has shown that the nocturnal Boreal Owl shows no apparent ability to see the scent marks of the voles it hunts. However, when Northern Pygmy Owls, which have more diurnal habits, were given the same test, they did appear to be able to detect the marks, while the Northern Saw-whet Owl's plumage fluoresces under ultraviolet light. Clearly at least some owls are able to see UV light.

SHORT-EARED OWL

TOUCHING MOMENTS

The sense of touch becomes important when an owl is closely engaged with something. Its long-sightedness means it cannot focus on what its own feet are doing, and in some situations its feet may be busy out of view anyway, for example when attacking prey that is under vegetation or snow. Specialised feathers called filo-plumes on the feet and around the bill act as sensitive touch organs, giving extra information about what is about to be seized or eaten. There are additional filo-plumes on the body, mixed in with, and hidden by, the larger contour feathers. They are also touch-sensitive and help the owl to arrange its feathers properly when preening.

SMELL AND TASTE

Most birds have little in the way of a sense of smell or taste, and owls are no exception. Their sense of taste, however, is at least developed enough for them to acquire aversions to eating animals that produce strong bitter-tasting chemicals, such as toads and certain species of moths. People with pet owls also report that individual birds will have clear preferences for certain types of meat over others, though this could be down to differences in texture as well as flavour.

EXTRASENSORY PERCEPTION

There is more to the sensory systems of owls and other living things than the five conventional senses of vision, hearing, smell, taste and touch. One of these 'extra' senses is the vestibular or balance sense, which we humans use to stop ourselves becoming dizzy and falling over. This is achieved by sensing where the head is relative to the body. The vestibular system is located in our heads, within our inner ears. It is composed of three looping tubular structures, called semicircular canals, which contain tiny rocky lumps called 'otoliths' (literally 'ear rocks'). The semicircular canals sense

The Great Grey Owl's stiff-edged facial disk helps channel sound to the ears.

Left: Northern Hawk Owl on high; many owls scan for prey from an elevated perch.

The how and why of bird intelligence

Defining and measuring intelligence is difficult enough when your subjects are your fellow human beings. With other animals, which cannot communicate easily with us and have evolved with very different lifestyles, it is necessary to be flexible and open-minded about what intelligence is, and how to measure it.

Brain size relative to overall body size is a useful indicator of intelligence. So too is brain complexity – the number of neurons and the size of the outer layer of the forebrain (the cerebral cortex). In mammals, you can also look at the extent to which the cerebral cortex is convoluted or folded (allowing more neurons to fit into a smaller space), but birds lack these convolutions, even though their brains may be proportionately larger than similar-sized mammals. The functions of the cerebral cortex include those that we associate most with intelligence – consciousness, thought, memory, perception and attention.

However, studying anatomy only tells us so much, and it is observation of behaviour that really reveals how much processing power an animal possesses. Many animals perform complex behaviours, but if they are innate or 'hard-wired' from birth, they are not deemed a product of true intelligence. Therefore, one of the key indicators of intelligence is the ability to innovate, performing experimental new behaviours to solve problems or exploit new opportunities. One famous example from the world of birds concerns the wild Rooks that learned how to winch up the bags in refuse bins at a service station, gripping the folds of plastic with their feet as they went, until they brought the bin's contents within bill-reach. Captive Rooks have shown similar innovative skills, bending wires into hooks then using them to retrieve food from narrow containers. Other aspects of intelligence include the ability to learn new behaviours, the use of exploratory play and the ability to understand abstract concepts, all of which have been documented in captive birds.

Right: Long-eared Owl on high alert; a flexible neck makes up for the immobility of owl eyes.

when we turn or tilt our heads, and the movement of the otoliths lets us sense when we're moving forwards or backwards in a straight line. Our brains then pull this information together and send out signals in response – to control our eye movements so that we don't get dizzy, and to adjust the muscles in our legs and torso to keep us upright.

One of the popular 'animals doing funny things' videos on the internet is a clip of a Great Horned Owl and a NASA scientist. The owl is wearing what looks like a straitjacket, and the scientist is holding its body and moving it around, tilting and turning it in various directions. No matter how much its body is tipped and along which axis, the owl's head remains absolutely steady, compensating easily for the body movement. This is the work of its vestibular system, which is proportionately much larger and more developed than that of a human.

It is easy to see why an owl needs a superlative vestibular system when you watch one in hunting flight. Navigating through three dimensions, with changing terrain obstacles to negotiate and variable wind speed to deal with, requires considerable agility in the air, but the owl's essential sensory prey-finding kit, its eyes and ears, need a consistent input as it moves along. There-

fore, the owl's head needs to be kept under close and stable control at all times.

THE WISE OLD OWL?

Traditionally, humans are no great respecters of avian intelligence, as indicated by popular expressions like 'bird brain'. Modern research is changing this view, though, with evidence coming to light that some birds can innovatively come up with problem-solving skills, form abstract concepts, show self-awareness and use tools – all indicators of impressive intelligence.

In folklore and fairy tale the owl is often portrayed as the brightest of birds, no doubt because of its apparently large head and thoughtful, human-like expressions, and perhaps also its quiet and rather mysterious manner. However, strip an owl of its feathers and you will see that the head is not proportionately large at all, just very heavily feathered, with feather structure providing its unique shape as well as its bulk. The skull of an owl looks quite at odds with the feathered bird, as the flat-faced look is gone and the bill projects well beyond the eye sockets. The skull is dominated by those large eye sockets, and their rings of sclerotic bone, making the actual cranium look rather small.

Nevertheless, owls are definitely nearer the top of

the scale than the bottom when it comes to brain size relative to body size, an honour they share with other bird groups including parrots, crows, falcons and hornbills, and this does suggest that they are of fairly high intelligence compared to most other birds.

Relatively little research has so far been carried out investigating the cognitive powers of wild owls, with practical limitations on what can or can't be observed, but it is known that some owl mental abilities are truly exceptional. A study by Eric Knudsen and Masakazu Konishi in 1978 found that Barn Owls form a remarkably detailed auditory map in their brains, with sounds coming from different directions triggering activity in specific corresponding neurons (brain cells). Barn Owls have also been shown to have a very accurate memory for sounds and their locations. Whether this can really be defined as intelligence, though, is debatable.

Owls' nocturnal habits and tendency to be shy and secretive are some of the obvious practical difficulties that make it hard to study their intelligence in the wild.

There is, however, plenty of anecdotal evidence from people with pet owls that these birds are clever and quick learners. Bernd Heinrich, a professor of zoology at the University of Vermont, documented his interactions with a hand-reared Great Horned Owl, saying that the bird learned how to wake its owner for an early meal, would gently take food from his fingers without ever mistaking the digits themselves for food, and would play energetically with inanimate objects for hours.

One interesting example of innovative behaviour concerns the Short-eared Owls of Genovesa Island in the Galapagos. On this island – but not elsewhere in the Galapagos – the owls have learned to exploit the numerous storm-petrels that nest in tunnels and rock crevices on the island. The petrels are quick and agile in the air but can be caught in the brief moments that they are on the ground making their way to or from their underground nests. The owls hide behind rocky outcrops, creep up on their prey and even tuck themselves into the larger crevices, ready to catch a petrel as it arrives.

Left: The Northern Saw-whet Owl is a clever and adaptable hunter, able to catch prey larger than itself.

Learning all the small details of its territory helps the Ural Owl refine its hunting skills.

Hunting and diet

The majority of birds are predators of some kind, even if only part-time, in the sense that they catch and eat other animals. Robins uproot worms, warblers catch flies, even ducks and geese may catch the occasional dragonfly or baby frog to supplement their mainly vegetarian diet. However, hawks and owls are considered separately as the true 'birds of prey' because for most species, warm-blooded vertebrate prey forms a major part of their diet. Catching larger, stronger, more intelligent prey is generally much more challenging than feeding on insects, and owls exhibit an array of adaptations and specialisations related to hunting.

STYLES OF HUNTING

For an owl, hunting is a matter of life and death. As soon as its parents stop supplying food to a youngster, it must regularly hunt for itself to survive, and every failed hunt is energy lost, making it even more important that success comes next time. Learning how to hunt takes time – picking up worms or slow-moving insects is relatively simple but 'low yield'; catching a substantial meal is far more difficult. Only those that are skilled enough will survive the rigours of their first winter and, in due course, have the opportunity to breed. Some owls stick to a single main hunting method, while others are more versatile.

Stealth The main drawback with hunting relatively alert and mobile animals like birds and mammals is that they are always on the lookout for danger, and can and will take evasive action to avoid capture. They know when they are vulnerable and need to be extra careful, and they know where to go to be relatively safe from predators. So avoiding detection is a top priority for a bird of prey.

Many raptors search for prey in flight, at a considerable height, scanning the ground with a remarkably powerful gaze. Hunting rather short-sighted ground mammals, they are essentially undetectable to their quarry until the final strike, which is (hopefully) too quick to allow time to escape. Owls, even diurnal species of open countryside, are not much given to soaring and gliding flight when hunting (though may use it when travelling long distances), but instead often perform what is known as 'quartering' – a methodical low-level flight in a series of straight lines across a chosen space. The low-level flight means there is a risk they will be heard, if not seen, by the voles and mice they are searching for, but owls are among the quietest of all flying things.

Patience Hunting from flight is not always the best tactic. An alternative is to sit, wait and hope the prey comes to you. Many owls will practise this sit-and-wait hunting method, choosing an elevated perch and scanning the ground below (with both eyes and ears) for prey movement. Fence posts, stumps, branches and boulders are all suitable. In habitats with quite thick ground vegetation the chosen perch is between one

Target acquired – a Barred Owl in mid-swoop.

Left: Eyes on the prize – a Northern Hawk Owl watches carefully for prey.

Soft wings

Silence in flight is important, not just to avoid being heard by prey, but for a bird that relies so much on its hearing, the sound of beating wings would interfere considerably with the detection of moving prey. Most birds generate a swishing sound as they beat their wings, the air rushing against and between their long primary and secondary flight feathers. In most owls, both flight and body feathers come with built-in mufflers, with a soft, velvety layer of filaments on the surface to deaden sound. Additionally, the leading primary feather has a serrated outer edge, looking like a narrow comb, which further reduces sound.

These feather modifications mean that a Barn Owl, for example, can use a slow hunting flight that involves a great deal of flapping and even hovering while it scours the ground. This it has to do, for flying close to the ground means the owl cannot take advantage of the thermals (rising warm air currents) that enable raptors like buzzards to glide and soar at great heights with minimal flapping effort. However, there is some assistance from the air – flying into the wind allows the owl to exploit the uplift and save its wings some effort. This trait is a boon to the owl-watcher – stand with the breeze behind you and there is a fair chance the hunting owl will fly towards you.

URAL OWL

and two metres high. However, where visibility is better – for example, over snow-covered ground, a higher perch may be better. Species that prey on birds are also likely to opt for higher vantage points – the Eurasian Pygmy Owl and Northern Hawk Owl *Surnia ulula*, both of which take a lot of avian prey, are often seen perched at the very top of a tall spruce tree.

The fishing owls form a special category of sit-and-wait hunters. They often choose a perch very close to the water's surface or may even stand in the shallows. Individual birds have favourite fishing spots, which they will use night after night.

An owl may sit on a single perch for long spells or move frequently from one to the next. Owls that prefer hunting in flight will resort to the sit-and-wait method when conditions are not suitable for controlled flying – in strong winds, for example. Species that prey mainly on ground insects generally favour this hunting style over flight hunting.

Pursuit This kind of hunting describes a chase when the intended victim has seen the hunter and is actively trying to avoid being caught. To pursue prey successfully in flight, the owl has to be quicker than its prey either in straight lines or on the turn – or both. Most forest owls have short, broad wings, which make them adept at quick twists and turns – necessary to navigate their 'cluttered' environments, but also useful for chasing birds (especially in low light, when the owl's

superior senses give it the advantage over its prey). Even so, the chase is usually short.

Open-country pursuits are a real challenge. The larger, more powerful owl species can chase down running mammal prey over open terrain – Snowy Owls, for example, can successfully hunt Arctic Hares. Their chances, however, are much increased if they can make a surprise attack – a running hare accelerates fast and can make much quicker turns than the flying owl.

Owls may also give chase to prey on the ground. This seems unlikely on the face of it – those bulky, long-clawed feet hardly look suitable for running. But owls are deceptively long-legged and can use their wings to help carry them along as they hop and bound after prey. However, a chase after a small mammal on foot is only likely to succeed if it is short and the victim is somehow unable to run to a safe patch of cover. Insects and other invertebrates are easier to catch on the ground. Fishing owls may also chase through the shallows after slower-moving prey such as crayfish.

Ambush A more developed form of sit-and-wait hunting: a predator waits in ambush having chosen a hidden spot where a particular prey type is more or less certain to come very close by. The predator then has only a short distance to cover when it makes its strike, which must nevertheless be very quick.

Some owls join other avian predators to loiter by large bat roosts at dusk, ready to grab a meal as the bats come pouring out into the night. As mentioned in the previous chapter, the Short-eared Owls on Genovesa Island in the Galapagos have learned to ambush storm-petrels by waiting inside the rock crevices where the petrels nest. As this behaviour is not seen elsewhere (although Short-eared Owls and petrels do coexist on other islands), it seems to be a learned and culturally transmitted pattern rather than part of the owl's instinctual repertoire.

Flushing out This unusual hunting behaviour has been observed in Barn and Tawny Owls, among other species, normally at dusk. The hunting owl flies low and close along a hedgerow where birds are feeding or roosting, ready to grab one if it is alarmed enough to break cover. Sometimes the owl will actually strike the hedge with its wing-tips.

Theft Stealing prey from another predator is called kleptoparasitism. In some bird groups, notably the skuas and the frigatebirds, it is practically a full-time way of life. In owls and raptors it is less frequent but still a regular occurrence. The would-be thief chases and makes repeated dives at its victim, and may even seize the prey or the victim's feet in mid-air. If the attacker is sufficiently skilled and determined, it will eventually force the victim to surrender the kill.

Short-eared Owls have been observed stealing prey from the harriers that share their moorland breeding grounds, while Snowy Owls have been seen relieving Short-eared Owls of their kills. Snowy Owls have also been observed to successfully take prey from Arctic Foxes – a risky strategy. Owls are, however, victims of kleptoparasites as often or more often than they are the

Above: **Great Grey Owls can hear prey moving under thick undergrowth or even snow.**

A dangerous habit

The owl habit of sitting on posts has been fully exploited by those with an anti-owl and raptor agenda. Leg-holding gin traps positioned on top of posts have been responsible for many owl deaths over the years and they are still used illegally today. These so-called pole traps are a particularly barbaric and indiscriminate method of harming and killing birds – their use has been outlawed in England since 1904 but there have been several cases of owls killed by these traps between 2000 and 2010.

BARRED OWL

aggressors. Kestrels readily and successfully challenge Barn and Short-eared Owls for their prey, as do Hen and Northern Harriers. Owls hunting in the open and by day are clearly particularly vulnerable to these attacks.

THE CATCH

Owls usually seize their prey in their feet. Exceptions include very small or helpless prey, which may be picked straight up in the bill. But the usual strike is feet-first, with the talons doing the work of controlling and killing the prey. Whether the owl is gliding downwards from a hunting flight, dropping from a hover or pouncing from a perch, it extends its surprisingly long legs as it gets within range. Often the prey is not actually visible even at this point, so the owl is relying on detailed information from its ears and stimulation of the sensitive hairs on the feet to know when to seize.

Smaller prey may be snatched up and carried off in one movement, while bigger and heavier prey will often force the owl to land and spend some time dealing with it on the ground. Very big and strong prey animals may tumble the owl over and force a prolonged struggle, during which the owl strives to get and maintain a strong

Above and right: With its long tail and wide wings, the Northern Hawk Owl is agile in flight, able to chase birds or pounce on rodents with equal ease.

talon grip on the prey while leaning back to keep its head well away from danger.

As soon as the feet make contact, the toes squeeze tightly together, forcing the talons into the prey's body. If the owl's victim is a small animal, this is often enough to quickly kill it, but the owl may also need to deploy its bill, biting the head or neck area. When it does this it closes its eyes, to protect them from any damage that might be inflicted by a kick or bite from the struggling prey.

DIET

Anything that can be caught and killed is potential prey for an owl. As solitary hunters, owls have limits to the absolute size of prey they can take, but many are capable of handling animals close to their own size. Their diet may also be occasionally supplemented with carrion or even vegetable matter.

Mammals Most owls take some mammal prey and for some species the diet is almost entirely composed of a few or even just one species of small rodent. These mammals are mainly active at night and move around through grass, other vegetation or leaf-litter on the ground where they are difficult to see but easy to hear. When it comes to detecting danger, mice and voles have keen noses and good hearing but poor eyesight and are not able to scan any distance around or above themselves for predators. Many of the most obvious owl traits – silent flight, acute hearing, nocturnal habits – are adaptations that help them to hunt this particular kind of prey.

Small rodents make the perfect basis for an owl's diet, as they are abundant in most habitats. Voles are especially important; mice are generally less popular as they are quicker and livelier than voles. In the Arctic, lemmings outnumber other small rodents, and fluctuations in lemming numbers influence the populations of many other Arctic mammals and birds, creating a sort of lemming-based ecological economy. In a poor lemming year, ground-nesting birds suffer increased predation of their chicks from species like Arctic Foxes and Snowy Owls, while a lemming 'boom' year will mean a breeding bonanza for the hunters.

Shrews are also on the owl menu. These small mammals are not rodents but insectivores, and tend to be rejected or ignored by mammalian predators, who

Blakiston's Fish Owl hunts along rivers, chasing fish in the shallows or just waiting quietly for one to swim by.

find them distasteful. Owls, which have a limited sense of taste and in any case usually swallow small prey whole, have no such aversion. Bat remains are also found in the pellets of woodland owls. Barn Owls have been observed taking bats on the wing, though a bat's agility in flight would usually keep it out of reach of most owls. Woodland owls may sometimes catch bats at their roosts.

The larger the owl, the larger the potential prey. The bigger eagle owl species (genus *Bubo*) can tackle squirrels, rabbits, hares, mustelids and hedgehogs, and the young of small deer, foxes and other larger mammals. Eagle owls that take up residence in or near towns and villages may upset local residents by preying on domestic cats and smaller dogs, and many a North American newspaper has carried stories of Great Horned Owls attacking toy-breed dogs, sometimes while their owners were standing just a couple of feet away.

Birds Among the raptors, there are many specialist bird-hunters. Falcons like the Merlin and Peregrine hunt birds in the open, with high-speed chases and stoops, while the true hawks (genus *Accipiter*) use tremendous manoeuvrability and surprise attacks to catch birds in woodland. Among owls, though, there are few truly specialist bird-hunters, but most species will prey opportunistically on birds. As most birds are diurnal, owls have the opportunity to catch them easily at night if they can locate roosting sites.

Bird prey is often caught on the ground after a long gliding descent, a similar method that owls use to catch mammals. Ground-dwelling birds like pheasants, quails and other game birds, which are more inclined to run than to fly, are particularly vulnerable to owl attacks, as are small birds like finches and pipits that often feed on the ground. By night, birds of all kinds that roost in the open rather than in cavities may fall victim

Above left: Voles are numerous in many parts of the world and are a vital component of many owls' diets, including the Barn Owl.

Above right: Big insects can provide a substantial meal for an Elf Owl.

Left: The Northern Saw-whet Owl eats rodents but will also take larger mammals and birds.

Although it is one of the larger owl species, the Great Grey Owl only rarely takes prey larger than a vole.

to surprise attacks. However, in daylight conditions, most birds can outmanoeuvre or outfly owls in an aerial chase.

Owls may tackle birds considerably heavier than themselves. The Tawny Owl has been known to take Mallards, which are double its weight, while the little Burrowing Owl *Athene cunicularia* will take doves that are easily its equal in size.

Other animals The fish owls are specialist predators of aquatic life – fish and other water creatures such as frogs and crayfish. These they catch by pouncing from a perch low over the water or (with slower prey) by chasing through the shallows. Fish taken may be sizeable, as heavy or heavier than the owl itself, and will be carried or dragged away from the water before being eaten. Other owls may occasionally take small fish from shallow water, too, snatching them up without settling, while the fishing owls do not eat fish and water animals exclusively but also take land prey.

Open-country owls will hunt reptiles as readily as

they take small mammals, and use similar techniques to find and catch them. A snake can be challenging prey – the owl will seek to get a strong talon grip on the head to subdue it. The largest owls will even tackle young crocodilians.

Many owls take land invertebrates, and the smaller species may live mainly on this category of prey. Larger and slower-moving insects are more likely to be taken than fast-flying species – crickets, beetles and heavy-bodied moths are particular favourites. Other invertebrates such as earthworms and centipedes are also taken. Insects may be dismembered before consumption, with large and indigestible parts, such as legs, wings and wing cases, being discarded.

Carrion Very few birds of prey will shun an easy free meal in the form of carrion, especially when times are hard. Young owls in their first winter may eat quite a high proportion of carrion, as they are yet to establish a productive hunting territory of their own, and are also less skilled hunters struggling to survive at a time when

Owls preying on owls

One of the more surprising things about owl hunting and feeding behaviour is the extent to which they prey on each other, and on the diurnal raptors. This phenomenon, known as 'intraguild predation', is widely documented in owls. The European Eagle Owl has been known to prey on every other owl species that occurs within its geographic range, even the Snowy Owl which is close to its own size. It also takes large raptors including Peregrine Falcon and Rough-legged Buzzard, making surprise attacks on these skilled flyers at their roost. It is a similar story with the Great Horned Owl in North America. In England, the Tawny Owl is the largest widespread species and will hunt the four smaller owls, especially the Long-eared Owl, which has the most similar habitat requirements to the Tawny and so comes into contact with it more often.

Is this purely opportunisitic predation or do owls make a point of seeking out and killing each other? Other owls and raptors may form up to 5 per cent of the Eurasian Eagle Owl's diet, a much larger proportion than would be expected by chance, and bird of prey species may be virtually eliminated from Eurasian Eagle Owl territories. In England, the Tawny Owl is the dominant species of deciduous woodland with Long-eared Owls restricted to more marginal habitats, but in Ireland, where there are no Tawnies, there is a much denser population of Long-eareds, in all woodland types. So there is evidence that owl-on-owl or owl-on-raptor predation includes an element of removing competition, as well as hunting for food.

These encounters are by no means low risk for the attacker, so the rewards must be high. Great Horned Owls will take Goshawk chicks and are capable of killing adult Goshawks. However, adult Goshawks will respond fiercely to a Great Horned Owl in their territories and often succeed in driving it away – indeed, scientists use tethered Great Horned Owls as lures to trap Goshawks for research purposes. An aerial battle observed between a Great Horned Owl and a Red-shouldered Hawk ended in the death of both birds.

A Boreal Owl with its Northern Pygmy Owl prey – even the smallest species practise intraguild predation.

finding living prey becomes more difficult. Road kill is an obvious source of carrion, though owls themselves are vulnerable to being killed by a car when taking it. Barred Owls have been filmed visiting deer carcasses, even when the meat has begun to decay, and Blakiston's Fish Owls in Russia have been caught in traps baited with raw meat, set with the intention of catching carnivorous mammals.

Non-animal food Owls can fulfil all their nutritional needs from a diet composed exclusively of animals, and there are very few records of them consuming vegetable matter. There have, however, been documented occurrences of Burrowing Owls consuming cactus fruit and of Powerful Owls *Ninox strenua* in Australia eating figs.

A Northern Hawk Owl with prey – unlike the diurnal raptors, owls often carry prey in their bills.

DEALING WITH PREY

Depending on what has been captured and the owl's particular circumstances, prey may be eaten on the ground where it was caught, or carried away to a safer place for consumption. When carrying prey in flight, owls normally grip the prey in one or both feet, or in the bill.

Owls, in common with raptors, are protective of their prey, and with good reason. It is quite common for other birds to try to steal from owls. To help prevent this, owls will 'mantle' over their prey, concealing it with half-spread wings.

Small mammal prey requires little 'management' before being eaten. Once the prey is dead, the owl will usually manoeuvre it into the bill by gripping the head or neck, leaving the body dangling down, and will swallow

it head first. Anything that won't fit down the gullet, though, must be torn up and eaten in pieces, and some prey may require processing first – birds are at least partially plucked and invertebrates' tough and spiky parts removed. However, owls do still eat plenty of indigestible material, most of which does not get very far along the digestive tract before being expelled in the form of pellets. Consuming and then regurgitating unwanted items seems to be necessary for survival, from as early as the first week in an owl's life. Captive owls fed on raw meat without any 'roughage' will sicken and die.

Pellets Many different birds that eat animal prey produce pellets of undigested food remains, including waders, herons and some passerines. Some birds, such as Rooks, also produce pellets containing undigested plant material. However, it is the birds of prey and especially the owls that are best known for pellet production.

The gastrointestinal anatomy of the owl is rather different to that of most birds. Typically, birds have a crop – a muscular pouch leading from the oesophagus or gullet, which is used as a storage organ rather than for digestion. Birds that feed their young on regurgitated food store it in the crop, and it allows birds like storks or raptors to quickly consume a very large meal in one go, to be digested slowly later on. But owls do not have a crop, so swallowed food goes straight to the proventriculus or first part of the stomach, where digestive enzymes are secreted to begin the digestion process.

Next stop along the digestive tract is the second part of the stomach, the muscular gizzard, where vigorous contractions help break the food down. This is

where pellets are formed, as the digestible parts of the meal move on to the small intestine, leaving the rest – fur, feathers, bones and other hard parts – behind. These remains are compacted into a pellet, which the owl regurgitates. It is often said that pellets are 'coughed up' but this is a little misleading as we usually understand a cough to originate from the lungs! The whole process from eating to pellet regurgitation takes 10 or more hours.

Little Owls often chase insect prey on foot, and look on the ground for worms on the soil surface.

Studying pellets

Because owls swallow much of their prey whole, and because their digestive acids are rather less powerful than those of most birds, the contents of their pellets are often very well preserved. It is common to find intact skulls and other identifiable bones from their vertebrate prey within a pellet, and owl-pellet dissection has long been a popular 'nature studies' activity for schoolchildren.

Most owls are very much creatures of habit and will produce pellets at the same time and place every day, leading to sizeable accumulations. Finding such a 'stash' of pellets enables researchers to make a very complete study of owl feeding habits, in terms of both the species taken and what proportion they make up of the diet. Even prey that is eaten in pieces is often identifiable from pellet remains, if fur or feathers are examined carefully.

OWL PELLET

A place to live

An owl must satisfy all of its needs from its environment. At the very least it needs to be able to find catchable prey, water to drink and a safe place to roost or shelter from bad weather – what this means in terms of actual habitat varies from species to species. In order to breed, it will also need a suitable nest-site and enough prey around to feed not only itself but its brood as well. Finding a suitable territory is perhaps the most vital single undertaking in an owl's life, and defending that territory from would-be usurpers is a task that never ends.

OWL HABITATS

Owls are present on every continent except Antarctica, and make use of almost every kind of land habitat imaginable, from the frozen Arctic tundra to semi-desert on the equator. As a group, they are perhaps most closely associated with woodland and make use of every kind – rainforest, boreal pinewoods and all sorts of temperate woodland, but there are many open-country owls, too, and others that nest in woodland but do most of their hunting over open ground adjacent to the forest. Some owls even live closely alongside humans in towns and cities.

Woodland Owls that breed and hunt in woodlands conform most closely to the 'owl archetype'. They are nearly all strictly nocturnal, have superb hearing and low-light vision, are short-winged for manoeuvrability in the 'cluttered' woodland environment and have intricately patterned brown and grey plumage for camouflage. The genus *Strix* contains some of the most specialised woodland owls in the northern hemisphere, such as the Tawny and Ural Owls *Strix uralensis* in Eurasia, and the Barred and Spotted Owls *S. occidentalis* in North America. The Boreal Owl, a smaller species of the northern pinewoods on both sides of the Atlantic, is one of the most nocturnal of all owls and shows the most marked ear asymmetry of any owl species.

Woodland owls usually nest in tree cavities, whether natural hollows formed in older trees by partial decay or, in the case of some smaller species, the old excavations made by woodpeckers. Availability of suitable nest-holes can be a limiting resource on population density, and unsympathetic woodland management may make things worse, by removing the old trees. Providing nestboxes can help – a sizeable proportion of southern Sweden's Ural Owl population now uses nestboxes. Other possible nest-sites include the old nests of large birds such as crows or herons.

Boreal forest or taiga, often considered to be rather lacking in biodiversity compared to more temperate woodland, supports many owl species. The greatest diversity of owls in Europe (10 species) is found at approximately 57 degrees north, a latitude that passes through the Scottish Highlands, Denmark and southern Sweden, areas with a high proportion of pine forest.

The Northern Pygmy Owl is a bird of wild pine forests in northern latitudes.

Left: Concealment is key for nocturnal species like Long-eared Owls. Their habitat must include safe daytime roosting spots.

Owls of this kind of habitat are generally grey-plumaged, which provides good camouflage against the greyish pine and spruce trunks. The Tawny Owl, which occurs in brown and grey morphs, tends to be greyer the further north and east you go through its range.

Temperate broadleaved woodland throughout the world is also a productive owl habitat. Some of the species found in the boreal coniferous forest extend their range into this kind of woodland, while the *Strix* owls are specialists of broadleaved forest. Many of the more generalist owl species will use the edges of deciduous woodland where it adjoins open countryside.

Tropical rainforest is the most species-rich habitat on land, but the owls are not especially strongly represented in this kind of ecosystem. Some specialist rainforest owls include the Spectacled Owl *Pulsatrix perspicillata* of Central and South America, the Greater Sooty Owl *Tyto tenebricosa* of Australasia and the Maned Owl *Jubula lettii* of tropical Africa.

Open country Owls that live and hunt mainly in open country tend to look and behave a little differently to forest owls – in general they are longer-winged, paler in colour and less sedentary in their ways. They are also more likely to be diurnal.

Reasonably wild open countryside, whether meadow, grassy steppe, semi-desert, moorland or alpine tundra, supports high numbers of small rodents and invertebrates and offers excellent hunting through the warmer months. It may be a very different story in winter, though, with the prey species becoming much less active and more elusive, which is why some northern owls are strongly migratory, heading south in winter in pursuit of better food supplies. The same goes for species that breed at high elevations – winter condi-

tions in the mountains may force them to move down the slopes. This is called altitudinal migration.

The Short-eared Owl is an example of a species that sometimes follows a nomadic way of life, moving freely over considerable distances in response to changes in vole abundance, and breeding as and when conditions seem optimal. Species like the Barn Owl, which favours open countryside but is mainly sedentary, can suffer high mortality in very severe winters.

Some open-country owls nest in cavities. In northern Europe the Little Owl *Athene noctua* usually nests in a tree-hole, so most Little Owl territories adjoin woodland of some sort, even though the birds hunt primarily in open country. Some pairs find suitable nest-sites within derelict farm buildings. However, the Short-eared Owl nests in a scrape on the ground, needing only some shelter and concealment from ground vegetation.

STAKING A CLAIM

The area of land that a bird uses is defined by two key terms – home range and territory. The home range is the entire area that the bird uses, while the territory is the area that it defends from intruders of its own species. The extents of both vary dramatically from species to species. For a pair of finches, the territory and home range may be essentially the same throughout the breeding season, but in winter the territory effectively ceases to exist and the home range becomes much larger, as the birds flock with their own species and roam widely, using teamwork to find food. For colonial seabirds, such as Gannets, the home range may be huge, encompassing vast tracts of sea, but the defended territory extends to just a few centimetres around the nest.

So, while every bird that is not actively migrating has a home range, the existence and size of a territory

Still in the woods

The woodland owls of the temperate regions include some exceptionally sedentary species. British Tawny Owls rarely wander more than 5km away from where they were hatched, and once a pair has established a territory they will remain there for their whole lives, unless forced out by a successful challenger. It will often be possible to see the birds roosting in the same tree every day throughout the year. Woodlands offer shelter from the worst winter weather, which benefits both the owls and their prey.

There are many good reasons why these owls are better off staying put. Good territories may not be easy to come by, with suitable nest-sites (usually large tree hollows) often in short supply. Also, thorough familiarity with every detail of its environment can only improve an owl's hunting efficiency. Only times of severe prey shortage would prompt a British Tawny Owl to leave an established territory, though owls of more northerly forests, such as Boreal, regularly undertake cold-weather dispersal.

TAWNY OWL

Overlap and competition

What if two different species of owls live in the same habitat and make the same use of the same resources? The idea of niche separation says this should not happen. Natural selection pushes different species to use different resources, or to use them in different ways – the less competition, the more chance of survival and successful breeding. In reality, though, there is often some overlap between two or more species. As discussed in chapter 3, the high incidence of owls preying on other owls is probably due to competition for resources.

Where different owl species are separated geographically, they may use extremely similar niches, but if they share a habitat they may be forced to use narrower niches. A study on Long-eared and Tawny Owls in Sweden found that where the two species lived in neighbouring territories, they tended to take different prey, but where they lived some distance apart there was much greater overlap in their diets. The Long-eared Owls also bred less successfully where they lived alongside the Tawnies, suggesting that they were the losers in this particular competitive situation.

When rapid changes to ecosystems take place, owl species that were formerly separated may come into contact with each other, with dramatic results. The Barred Owl in North America has spread westwards from its traditional east Pacific range since the mid 20th century, probably because of man-made changes to the environment. This has brought its range into that of a subspecies of the near-threatened Spotted Owl. The two species are closely related and ecologically very similar, but the larger Barred Owl has proved capable of outcompeting the Spotted and has displaced it completely from some areas.

Adapted to a life in semi-desert, Burrowing Owls rarely come into competition with other owl species.

For summer or for ever?

Pairs of sedentary owls usually defend a sizeable territory around their nest-site all year round, within a larger home range which may overlap with that of neighbouring owls. This year-round territoriality is not surprising, as prey availability and feeding behaviour do not change very dramatically from season to season. Also, with competition for nest-sites being as significant, if not more so, than competition for food, keeping hold of a prime nesting spot through the winter is vitally important, especially with a new crop of young owls looking to establish territories.

For owls that migrate south for winter, the breeding territory must be abandoned. On returning from migration the owl will usually seek to reclaim its territory from the previous year, or try to 'upgrade' to a better one. Migrants that return earlier have more chance of success in this respect, but run the risk of being on territory before there is sufficient prey around.

Migratory owls may establish feeding territories on their wintering grounds. Whether they do so depends on prey abundance – if there is a real glut of prey, the energy invested in trying to chase off other owls from a territory would be much better spent just hunting (and perhaps caching surplus prey for later).

SNOWY OWL

depend on what resources it needs, and how scattered and abundant those resources are. These factors determine whether the resource is worth the energy required to defend it. For a Gannet, it would be highly impractical to attempt to defend a patch of open sea, far from the nesting grounds, as a feeding territory for itself alone. For our finches, defending a feeding territory immediately around the nest is not so difficult, but in winter when their diet changes from mainly insects to mainly seeds and fruits, food resources become more 'clumped', and so it makes more sense to wander further and seek food in a flock.

Territorial behaviour Birdsong, often quite beautiful to our ears, carries a stern message of warning to other birds of the same sex and species – stay away, this patch is taken. Most birdsong is performed by male birds, and as well as discouraging other males it is used at the start of the breeding season to attract a mate. While the best-known owl vocalisations don't really match our idea of birdsong, they have the same functions. Among owls, attracting a mate appears to be a more important song function than territorial defence.

Nocturnal owls are the most vocal, and their voices carry considerable distances through the otherwise quiet night. Most owl songs are variations on a hooting, cooing or piping theme, following a predictable sequence of long and/or short notes. Larger species tend to have deeper voices – the eagle owls (genus *Bubo*) have quite booming songs while those of the tiny pygmy owls (genus *Glaucidium*) are shrill whistles. The barn owls (family Tytonidae) are quite different – their territorial songs are harsh, grating shrieks or screams. Songs are usually delivered nearer the territory boundaries than the centre. Some owls will sing ceaselessly for hours on end. Once the owl has paired up, singing becomes a much less frequent activity. While there may still be a need to deter intruders, the male doesn't have the time for regular singing, as he now has to hunt for his partner and, soon, their chicks. Of course, not every owl is lucky enough to find a partner, but unsuccessful males remain optimistic – unpaired male Common Scops Owls *Otus scops*, for example, may sing continuously each night from April to September. This may not be as pointless as it seems – perhaps females breeding in neighbouring territories will be paying attention and will pair with the relentlessly singing male the next year, especially if they fail to breed successfully with their current mate.

Another way of demonstrating territorial ownership is by a flying show. The diurnal raptors are masters of aerial displays, males and sometimes both members of a pair together performing dramatic roller-coaster flights high over their territories. They are visible from a considerable distance, and the display is an excellent means to show off their health and plumage condition – helping to discourage rivals and, for unmated birds, impress would-be partners. It also means that they can check and patrol their territory at the same time. Some of the day-flying owls, such as the Short-eared and Snowy Owls, exhibit aerial territorial displays.

NEST-SITES

Owls are not gifted nest-builders. The Short-eared Owl is one of the very few species that makes any attempt to make a nest, and that is limited to throwing a few strands of grass in a scraped-out hollow on the ground. This lack of creativity certainly saves time and energy, but because the owl cannot create a nest from scratch, it relies on suitable pre-existing sites, which are not always easy to find.

Short-eared Owls are adaptable nomads, able to hunt successfully over all kinds of open habitats.

Far left: Abandoned farm buildings are just as good as hollow trees for breeding Little Owls.

A song for two

One phenomenon seen among owls but not most other bird groups is that of duetting. This is a part of courtship and involves the pair singing together, timing their vocalisations so both 'parts' of the song are clear. Duetting is very common among owls, serving to cement the pair bond with species that pair for life, and to establish it among those that are only together for the breeding season.

BURROWING OWLS

Nestboxes

Owls very willingly take to nestboxes, providing they are suitably sited. Since the late 20th century, nestboxes suitable for the commoner suburban owls have been available in shops, and it is easy to find detailed specifications for boxes to suit practically any species. Supplying boxes helps to offset the loss of existing sites when, for example, old trees are felled or old buildings are patched up. For some species, providing nestboxes has made a great difference to breeding success and abundance. In Japan, nestboxes have allowed Blakiston's Fish Owls to continue to breed in areas where logging had removed long-used nest-sites. In Canada, elaborate artificial burrows are being provided for reintroduced populations of Burrowing Owls.

EASTERN SCREECH OWL – RED MORPH

Other birds' nests Some birds go to the trouble of constructing a nest from scratch but only use it for one season, starting again each year. The old nest, if it survives the winter intact, may be suitable for reappropriation by a pair of owls. Species that use old birds' nests include the Long-eared Owl and the fish owls. Suitable nests need to be big enough – for the huge Blakiston's Fish Owl only the nest of another very large bird such as a heron will do. Some small South American owls will reuse the enclosed, very solid, ball-shaped mud nests made by birds of the family Furnariidae or ovenbirds.

Tree-holes Cavities in tree trunks or large branches make excellent nest-sites – weather-proof, elevated out of reach of most mammalian predators and inconspicuous. Natural holes may form in trees when a branch breaks off and the subsequent wound becomes rotten, or if some kind of injury or disease within the tree causes it to decay from the inside.

The perfect tree-hole nest, however, is one made deliberately for that very purpose. Woodpeckers of various sizes occur throughout the world and have the equipment necessary to carve out a nest-hole in a tree trunk. They sometimes reuse nest-holes in successive years, but often with reduced success as over time climbing predators such as martens (weasel-like woodland mammals) will locate the holes, and in most woodlands it is easy to find numerous long-abandoned woodpecker nest-holes. These make perfect homes for owls of comparable size to the woodpecker species that originally made the hole. In some cases, a pair of owls may even evict woodpeckers from a brand new nest-hole.

Owls can also be vulnerable to exploratory invasions from martens but are generally better able to defend themselves. However, this kind of attack is the single biggest danger to owls that nest in tree-holes, especially the smaller species.

Other cavities The Burrowing Owl is well known for nesting in the burrows of tunnelling mammals, such as prairie dogs and ground squirrels. These tunnels offer shelter and a certain degree of concealment, but predators like American Badgers can easily access them. The owls have various tricks for improving their odds of success, from distraction displays to using smelly cow dung to mask the scent of the nest, but choosing any kind of hole or tunnel at ground level is fairly high risk.

A better option is a hole above ground, beyond climbing reach of marauding mammals. Holes in rock faces are used by many larger owls, and cracks in walls by smaller species. Barn Owls will nest on open ledges inside buildings as long as the access hole is small enough and well elevated.

Out in the open Species like Short-eared and Eastern Grass Owls *Tyto longimembris* that nest among vegetation on the ground are taking the biggest risk of all, though they benefit in that they have a much wider range of possible sites than species that insist on a cavity of some kind. This kind of ad hoc nesting arrangement makes most sense for species that wander nomadically and breed opportunistically – it enables them to set up home quickly when conditions are good for making a breeding attempt. Ground-nesters pick a spot that is not visible to the casual observer, tucked under an overhanging bush or stand of long grass.

Some owls will nest in fairly open and visible situations on cliff or rock ledges, out of reach of mammals. Owls that use these kinds of sites tend to be the larger

Some owls show plumage variation depending on habitat. Southerly subspecies of Little Owl (*above*) are paler and sandier than northern birds.

Left: A Northern Hawk Owl uses its grey plumage to camouflage the entrance to its nest-hole.

Right: Eurasian Eagle Owls often nest on the ground, but need sheltered spots to protect them from weather and nest predators.

species, more than capable of fending off winged nest robbers like crows.

OTHER HABITAT USES

The annual breeding season places heavy demands on a pair of owls. Without a secure nest-site and easy access to plenty of prey, all that hard work may well end in failure. At other times of year, things can seem easier but are not without their own challenges. After breeding comes the annual moult. Regrowing a full set of feathers is costly in energy terms, and mid-moult no bird is at its best – the ability to fly and keep warm may be impaired. Heading into winter, although the owl has fresh plumage, it faces possible savage weather and difficult hunting conditions, sometimes to the point where it will have to undertake a perilous migratory journey to find enough to eat. Wherever it is, it needs a safe place to roost, and it may still need to defend its food supply from intruders. Its habitat needs when not breeding may be different but are no less exacting.

In bad winters, Northern Hawk Owls roam widely to find feeding grounds and may turn up in towns.

Winter feeding territories Owls that leave their territories for winter will go in search of a new place to live and feed. True migrants, which move predictably from A to B for the winter, may have different needs to nomadic species, which wander around and may spend a day here, a week there or a month somewhere else. But birds in both categories may establish a territory and defend it from others. A wandering Snowy Owl may make a series of short stop-offs and defend a territory in each of them. If prey is very abundant in a particular location, an owl is better off feeding than expending its energy trying to drive off the many other owls that arrive, and in such situations several owls may be seen hunting in the same small area.

Roosts Because most owls are nocturnal, they cannot rely on the cover of darkness to keep them hidden when at their roosts. The hours of inactivity place an owl in a very vulnerable position. When other birds find a roosting owl, they will often mob it mercilessly, flying repeatedly at it while giving constant loud alarm calls that attract more mobbers to the scene. A dozen or more different species will join in, mainly small songbirds, and the owl stands little chance of sleeping when this noise and bombardment is going on. It will probably have to move on and is at increased risk from predators when on the move in broad daylight.

Predators may also directly target roosting owls. Most dangerous to owls are in fact other owls – virtually all owl species will kill other, smaller owls when they get the chance. Only the largest species in their respective

Woodland owls like Long-eared Owl have mottled brown plumage that provides good camouflage against tree bark.

Below: Snowy Owls blend into their surroundings beautifully through most of the year.

habitats are safe. Raptors will also take owls at their roosts. Mammalian predators, such as foxes and Coyotes, will seize the chance to take a roosting owl on the ground, and climbing species like martens may take owls roosting in trees.

Finding a safe roost site is therefore at a premium. The superb camouflage seen among most smaller owls is a defence against predators rather than a means of hiding from prey. The scops owls and screech owls have particularly impressive bark-like patterning on their plumage and ear-tufts, so that when they rest on a branch they are easily mistaken for the broken-off stump of a smaller branch. They may also rest close to the trunk, pressed against it in a sleek, tall posture that helps them blend in from all angles. Many owls that hunt in the open need trees around as much for roost sites as for nesting places.

Roosting in a hole or other kind of cavity or crevice offers concealment and protection because most predators would not be able to access the hiding owl. Catching the owl, however, may be very easy for some animals – if a weasel or marten enters a crevice where a small owl is roosting, there is little chance the bird will escape.

Drinking and bathing Owls do not drink very much water, as their diet supplies most of the moisture they need. However, access to fresh water is necessary for when they do need to drink, and many owls enjoy

bathing and will often drink a little during a bath.

Feather care is an important part of owl life. Flight feathers need to be 'zipped' back into shape if disarrayed or they will not function properly, and body plumage must be well maintained to help regulate body temperature. Daily preening is part of the care regime, and in many species pairs mutually preen each other's head and neck feathers – a bonding activity but a highly practical one, as these feathers are difficult for the owl to reach itself. Bathing helps clear dust and dried-on dirt from the feathers, including the more inaccessible ones.

To bathe, an owl needs shallow fresh water that is easy and safe to reach from dry land. Flattish gravelly riversides are ideal. The bird wades in about belly-deep and then performs vigorous dipping and thrashing movements, splashing water all over its plumage. Of course, it is vulnerable to predators while doing this and also when it has finished bathing and its plumage is still weighed down with water. Parts of owl feathers crumble to produce 'powder down', a fine dust that helps waterproof the feathers, so bathing owls don't become waterlogged enough to prevent flight. But an owl's ability to take off is certainly somewhat impaired after a bath and to minimise the risks, it will seek bathing spots that offer good all-round visibility and reasonably low branches that it can easily reach if a predatory mammal comes along.

Other kinds of bathing

As night birds, owls are not especially known for sunbathing, but may indulge occasionally. Sunbathing birds orient themselves so the sun hits their backs, fluff up their body plumage and spread out their wings and tail. One apparent function of sunbathing is to reduce the parasite load in the plumage, as the warmth apparently encourages feather lice to depart.

Anting is a peculiar behaviour whereby a bird allows ants to climb all over its plumage, sometimes even deliberately placing ants among its feathers. Its function is not fully understood but appears to be related to parasite control. The very few reported cases in owls suggest that access to ants' nests is unlikely to be an important requirement when selecting habitat. Owls occasionally dust-bathe in spots with dry loose soil, shaking the dust through their plumage as if bathing in water. As with bathing spots, good visibility is necessary for a suitable dust-bathing hollow.

Communal roosts

Usually, roosting alone is an owl's best bet. It is easier to hide one bird than many, after all. Even bonded pairs at the height of courtship will not necessarily roost together. However, roosting in groups does have the advantage of extra eyes and ears to detect approaching danger. Short-eared Owls usually roost on the ground, and while a single fox could quite easily find, stalk and kill many owls in one night if the birds live separately, any disturbance in a communal roost will quickly alert all the birds present. Another benefit of a communal roost is when it serves as an information centre. Observing the flight paths of arriving and departing owls provides each individual with information about where to go for productive new feeding areas.

LONG-EARED OWLS

Courtship and nesting

An owl's lifestyle is not generally much improved by company. An extra owl working the same hunting grounds means less food for both. Sharing a roost with another owl increases the chances of discovery by a flock of mobbing birds or, worse, predators. Complete solitude would seem to be ideal, but the most important goal in an owl's life is to breed, and for that it needs a partner. Owls devote great amounts of time and energy into attracting a mate, developing a pair bond and then together finding and defending a nest-site – all essential groundwork if they are to manage to rear chicks later on.

MATING SYSTEMS

Throughout the natural world, animals that reproduce sexually (rather than simply cloning themselves) have many different ways of pairing with the opposite sex. Even just within the bird world there are a variety of arrangements. In some species, such as most game-birds, the sexes only meet to copulate, with males seeking to pair with as many females as possible and females looking for the best-quality males.

It is more usual for males and females to stay together through the breeding season but the exact arrangement varies from species to species. In the Ostrich, for example, males guard a 'harem' of females, mating with them all and defending them from other males as well as protecting them and their broods from danger throughout the breeding season. In a few species, a single female pairs with two males for the duration of the breeding season, or vice versa. In the case of the Dunnock, both of these go on at the same time, resulting in very complex love quadrangles with both sexes having multiple mates.

The most common mating system among birds, though, is monogamy – one male and one female rearing a family together. The monogamy may be seasonal only, with the pair separating when the season ends and not rekindling their bond (unless by pure chance) the next year, or it may endure through the seasons and be a lifelong partnership. Which system a particular species uses depends largely on the way it feeds and how much care its chicks require.

One parent or two? For birds that produce precocial chicks, active and able to feed themselves hours after hatching, parental care is relatively simple – all that is required is to lead the chicks to suitable feeding grounds, and take whatever measures possible to hide or defend them from predators. Species that have precocial chicks often live in one-parent families, with just the mother (or sometimes the father – once the eggs are laid it makes little difference) incubating the eggs and caring for the chicks. However, solo incubation is problematic if the bird cannot quickly and easily leave the clutch and feed itself, and in cases like this the other parent will remain around during incubation at least, to warm and guard the eggs when the first bird is taking a break.

Day-flying owls like the Northern Hawk Owl often call and display in flight to attract a mate.

Left: The Boreal Owl is one of many species that nests in tree-holes, sheltered and safe from most predators.

Polygamy in owls

With regard to animal biology, polygamy usually means having more than one partner long term (rather than just for a single mating), without specifying sexes. If a male has two or more female mates, the term is polygyny, while the reverse is polyandry. All owls are normally monogamous, but both polygyny and polyandry have been observed on occasion in many species of owls.

For a male, the advantages and disadvantages of having two mates are clear. With two nests, he stands more chance of producing more chicks. However, because the male owl's role is to supply his mate and chicks with food for at least five weeks, this doubled workload is a distinct disadvantage. There is a good chance that some chicks will starve. Also, established females will usually not tolerate intruding females in their territory. Only a male with a very good territory in a productive year is likely to make a success of bigamy.

Much better, from the male's perspective, would be to stick with one partner but take any opportunity he gets to copulate with a paired-up neighbouring female, and thus father extra chicks but leave the hard work to another male. This is common among otherwise monogamous birds and probably does occur in owls, but while it greatly benefits the perpetrator, it spells wasted effort for the cuckolded male. Male owls minimise the risk of falling victim to cuckoldry by staying close to their mates throughout the courtship period. For females, cuckoldry is not particularly important either way.

A female owl with two mates would have a great advantage – with two suppliers for one nest, she and her chicks should be well provisioned. For a young male owl, seeking his first territory, sharing a mate with another male might be a better option than nothing at all, but the original male is unlikely to tolerate the interloper.

Polygyny has been recorded quite often in Short-eared Owls, a species whose nomadic habits make it more relaxed about territorial boundaries when a breeding opportunity presents itself. In owls that establish permanent territories it is rarer but has been documented in a range of species including Boreal, Burrowing, Snowy, Northern Hawk, Barn and Tawny Owls. Amazingly, a case of trigyny (one male with three mates) has been noted in the Northern Saw-whet Owl .

Comparable cases of polyandry are much fewer and further between. A variant of polyandry has been noted in Boreal, Northern Saw-whet and Long-eared Owls whereby a female has a brood with one male and then abandons the chicks to his care and has a second brood with a different partner, so-called serial polyandry, but simultaneous polyandry seems to be very rare.

Cases have been observed of extra adult Eastern Screech Owls *Megascops asio* and Burrowing Owls feeding a brood of chicks, possibly unrelated males, but there is also the possibility that they were older siblings of the brood they were tending. In one study a male Long-eared Owl that was helping a pair to rear a brood of chicks was found (by DNA testing) to be related to some of the chicks, but it could not be determined whether he was their father or brother.

In owls, one male paired with one female is the norm, as with these Collared Scops Owls, but other arrangements have occasionally been noted in some species.

In species where the chicks are altricial – rather helpless for some days or weeks after hatching and needing to be fed, childcare becomes a two-bird job. Unlike mammals, with a built-in food supply in the form of mammary glands, these birds must go out and fetch food for their young. This is why monogamy is so much more common among birds than mammals.

Parental roles in owls All owl species form pair bonds of seasonal duration at least. Because owl chicks are helpless when hatched and cannot feed themselves for many weeks, they need to have food brought to them. Also, an incubating owl cannot risk leaving its eggs alone when it needs to feed, because a successful hunt may take too long and the eggs could chill. The other parent must step in, either by taking over the incubation while the first bird hunts, or by bringing food to the incubating bird.

Owls all solve these problems in the same way. The female incubates the eggs by herself, and the male keeps her supplied with food. For the duration of the incubation period he must hunt for two. Meanwhile, the female can remain on the eggs and small chicks almost non-stop, and if any danger threatens the nest, she is

ready to do what she can to protect it. Only when the chicks are quite well grown, and no longer in need of her warmth, does she leave them alone. Now both parents have the same role – hunting constantly to keep the chicks fed.

Thus both parents have considerable, and distinct, responsibilities during the breeding season. The process of courtship between two owls is a way for both male and female to assess each other's potential ability to meet those responsibilities, and to show off their credentials as potential parents.

Mate choice For an unpaired female owl, selecting a mate goes hand in hand with choosing a territory. She responds to unpaired singing males by visiting them and their territories, and will quickly gain an impression of the quality of the territory and of the potential nest-sites it holds. The quality of the male himself is of less importance, although she may be able to evaluate his fitness before courtship even begins, by the quality of his song and, in some species, his display flights. During courtship, the male engages in courtship feeding, catching prey and offering it to the female. His ability to do this shows her that he is a capable provider.

A pair of Barred Owls indulge in long mutual preening sessions.

In species that do not form a pair bond, the female raising the young alone, mate choice is really only exercised by females. If the male's investment in the breeding season is limited to just the act of mating, his best tactic is simply to copulate with as many different females as he can. In this way he will father as many offspring as possible. The female invests far more time and energy into rearing just one batch of young at a time, so for her choosing the right father for them is much more important. This generally gives rise to a system whereby males compete among themselves for supremacy, and females mate only with the most successful. However, in monogamous species both sexes make about the same investment of time and energy into a brood, so, in theory at least, both should be choosy about their mates.

When females have a choice of males, they do exhibit certain preferences. In particular, they will prefer an older male to a yearling or two-year-old bird – experience is a valuable commodity when it comes to picking a breeding partner. But in general the older males will be occupying better territories than younger birds, making the choice straightforward. When it comes to male choice of females, though, often the first female that comes along will be accepted just because of time constraints. However, there is some evidence that when given a choice males have ways of comparing female health and selecting a mate on that basis. One study showed that male Barn Owls prefer females with stronger breast spotting, which has been shown to be correlated with better resistance to parasites.

Both sexes may also compete among themselves for mates. If an owl accepts a mate and that mate is then ousted from the territory by a rival of the same sex, the usurper will generally be accepted as a new mate, having shown itself to be strong and competitive. This

Reversed sexual size dimorphism

In most owl species, the females are larger than the males. The same goes for the raptors, some species of which have very marked differences – a female Sparrowhawk, for example, can weigh nearly twice as much as a male. The difference is not this marked in owls but a clear difference can be seen in most species. This reversed sexual size dimorphism or RSSD ('reversed' because it is more usual among birds for the male to be larger) has several possible explanations.

One theory is that two different-sized birds are better able to access a wider range of prey types, and compete less with each other. The small male is quicker and more agile, making him an effective hunter of small, fast prey, while the larger female can overpower prey that is too big for the male to handle. Another reason why bigger may be better in females is nest defence – the female is the one that spends most time on the nest so is most likely to have to deal with any predators that threaten it. A larger female may also be able to gain more weight prior to egg-laying, to lay more eggs and to be a more effective incubator, especially if there are lots of eggs in the clutch.

On the other hand, a small male may lose his territory to a larger rival. Studies have shown that territorial male Common Scops Owls' normal calls are influenced by their body size, with larger birds having deeper voices, and when challenged by a calling rival, they will 'put on' a lower voice.

There is some debate over whether mate choice among owls helps to maintain the size difference between the sexes – do males pair preferentially with larger females, and do females choose small males? A study on Barn Owls suggests that size is not a factor in mate choice, while work on Eastern Screech Owls has been inconclusive. If RSSD is not driven (or not wholly driven) by mate choice, then other factors must be at work to maintain the balance, suggesting that for whatever reason larger females and smaller males are better able to survive and/or reproduce.

LITTLE OWLS

Above: Mutual preening has practical benefits and helps affirm the bond between male and female, as with these Barn Owls.

Long-eared Owls form pair bonds that only last for the breeding season, but are quite sociable with their own kind at other times of year.

A pair of grey-morph Western Screech Owls stand guard in their nest tree.

has led to cases of hybridisation between Barred and Spotted Owls, with intruding male Barred Owls successfully driving male Spotted Owls from their territories and pairing with the females.

COURTSHIP

The rituals of courtship initially serve the function of allowing a prospective pair to size each other up, and to help both birds get ready for breeding. Owls in long-term relationships will go through a courtship phase at the start of each breeding season, characterised by several distinctive behaviours.

Song and dance Owls are well known for performing duets. An unpaired male sings, often for hours through the night, until he attracts a female, and she sings back as she approaches him. Once paired the two sing together, which helps deter intruders of both sexes and may also help the pair build an attachment to each other. The female's song is usually similar to the male's, sometimes shorter, simpler and higher pitched. The duet may involve both birds singing at once, their voices overlapping, or be timed so that the two birds' notes sound individually.

In diurnal owls that live in open habitat, display flights are another way to attract a mate. They are usually performed by the male, though in pairs the female may join in, and consist of ritualised flight patterns with particular exaggerated movements. The male may fly in a wide circle or with steep climbs and dives. He will often call or sing while flying and may wing-clap as well. Some flight manoeuvres show off striking parts of the plumage, like the pale underside of the wings in Short-eared Owls.

Courtship feeding The male owl offering the female an item of prey often precedes the pair's first copulation, and this ritual is repeated many times in the run-up to nesting. It is a clear demonstration of his hunting ability, and it also helps the female get into breeding condition – fatter females have more bodily resources for the energetically demanding process of forming eggs.

Mutual preening Mutual preening is documented for most well-studied species. This most touching aspect of owl courtship involves the pair sitting close together, usually in full side-by-side contact, and preening each other's head, neck and breast plumage. It is difficult to see a way that this activity directly relates to the parenting tasks ahead – perhaps the tenderness involved helps to demonstrate an ability to be gentle with small owlets. But it seems equally plausible that the owls do it as a service for one another, as it helps maintain good plumage condition in the places that are hard for an owl to preen by itself, and perhaps simply because it feels pleasant and helps build a feeling of closeness and trust between the pair. Those who own pet owls will testify to how much the bird seems to enjoy having its head and neck feathers gently scratched and stroked.

NESTING TIME

As discussed in chapter 4, owls do not build their own nests, so they need some kind of pre-existing site that can be used for a nest more or less as it is, whether this be a tree-hole, an old stick nest made by a different bird, a cliff ledge or even just a hollow on the ground. A male owl occupying a new territory will look for a patch that contains at least one and ideally more possible nest-sites. A good nest-site is the single most important element of an owl's territory, and the best ones will be used every year for decades, with new birds waiting in the wings to move into the desirable patch when one of the resident pair dies. However, if a pair of owls loses their brood to a predator they will generally avoid using that nest-site again. Most owl nest predators are long-lived and intelligent animals that will remember where they found a good meal and try the same spot again the next year.

When he attracts a mate, one of the male owl's first actions will be to show her the nest-sites he has found. He does this by going to each site and singing from next to or inside it. She approaches and explores the site, and may make second or third visits before making a selection.

Once she has picked a site, the female owl will begin to roost there and to spend time doing what little modifications and renovations she finds necessary. With a nest in a hole this usually amounts to nothing more than scraping out the bottom of the cavity to make a hollow for the new eggs. With old stick nests the modification may be even less than this. If the site is being reused from the previous year, there is likely to be a soft cushioning layer of last year's pellets and assorted other debris still there, and generally the female will leave most of this where it is, only removing what is directly in her way.

In the few days prior to egg-laying, the female gets almost all of her food brought to her by the male, and spends long hours in the nest in between frequent copulations. This spell of rest and plenty of food enables her to gain weight and reach peak condition for the long weeks of incubation ahead.

A male Great Grey Owl brings food to his incubating mate, so she need not leave the nest.

From egg to adult

Having established a territory, gone through courtship and selected a suitable nest, the owl pair now begin the drawn-out and challenging process of bringing the next generation into the world. This annual routine is a period of hard work for the parents, but they are driven by a biological need to pass on their genes, and if they are skilled parents and live long lives they could produce dozens of offspring. However, as soon as the young owls are independent of their parents they face perhaps an even bigger challenge before they too can breed – making it through their first winter alone. Only those with a good start in life will make it.

ABOUT EGGS

Because they need to be able to fly, birds have to lay smaller eggs than might otherwise be optimal, with longer intervals between the development and emer-gence of each egg. Carrying a larger egg or a number of fully formed eggs at the same time would weigh down a female owl too much and impede her ability to escape danger, risking her life as well as her young. Small eggs containing tiny embryos need a long time to develop, and constant warmth. Therefore at least one parent is committed to an incubation period that lasts three weeks or more, depending on the species. These are the evolutionary compromises that enabled birds to fly despite their heritage as warm-blooded egg-layers, and flight gave them the means to use much safer nesting places.

Bird eggs have hard, waterproof shells, so they don't lose moisture. Dietary calcium is needed for this. However, the shell is gas-permeable to enable the embryo to breathe. Thousands of tiny pores in the shell allow gas exchange.

A Snowy Owl mother feeds a young chick, with great care and gentleness.

Left: From a young age, owlets show great interest in other living things. This is a Northern Hawk Owl fledgling.

Egg formation

After copulation, the male's sperm is stored within the female's body in structures called sperm storage tubules until required (usually a matter of days). A female owl's reproductive cells – ova – are formed in her ovary (owls, like most birds, have only one developed and functional ovary and oviduct). A fully developed ovum, complete with a supply of yolk, is released into the oviduct, where it is fertilised with the stored sperm. Then it proceeds through the female owl's reproductive tract, gaining layers of albumen (egg white) as it goes, sealed in with membranes. Finally the calcium-rich shell layer is added, just before the egg is laid.

Incubation for this Great Grey Owl will take the best part of a month.

Egg appearance Owls exhibit dramatic variation in terms of how many eggs they lay (see 'Clutch size'). However, across the entire range of species there is great uniformity in the eggs' appearance (except for size). Every owl lays white eggs, and those within the 'typical owls' (family Strigidae) are also noticeably more rounded than most other bird eggs, almost spherical. White eggs are usually the mark of a bird that nests in cavities – there is no need for camouflaged markings or pigmentation on the eggshells if they are hidden in a hole.

As for the egg shape, a sphere has a maximal volume-to-surface-area ratio – space for the chick to develop but easy to keep warm. The longer egg shapes of other bird species are adaptations to particular needs – eggs that spin rather than roll or narrower eggs for birds whose manoeuvrability would be affected by rounder eggs. As they nest in hollows, owl's eggs have nowhere to roll to, and female owls do not forage for themselves in the run-up to egg-laying so can afford a little physical inconvenience. Across all owl species, size is the most obvious difference between their eggs.

Clutch size In nature, animals that suffer high predation pressure tend to have many more offspring than those nearer the top of the food chain. In this way the prey species offsets its inevitable losses. This can be seen among owls – the smaller species, which are preyed upon by other owls as well as various mammals, lay more eggs per clutch on average than the large species. There are also differences within species. Those that breed across a wide latitudinal range will generally produce larger clutches further north, reflecting the more challenging lifestyle they have, often involving winter migrations.

Even within a single individual pair, clutch size may be dramatically different between years. Several of the northern owl species have a heavy dependency on the local vole population, and in good vole years will lay much larger clutches to exploit the bounty. The Snowy Owl may lay as many as 15 eggs in a good year, but only two or three when vole numbers are low – or may not even attempt to breed at all. Species with a more consistent food supply show much less variation in clutch size.

INCUBATION

Once a pair of owls begin to mate, the female will spend a lot of her time in the nest, and when she has laid her first egg she begins to sit tight. She has already depended on her mate to bring her food in the days running up to the first egg being laid, but now she is completely reliant on him, as she will do all of the incubation and will rarely if ever leave the nest until the eggs have hatched and the chicks are a couple of weeks old. The male does no incubation and in fact will probably not enter the nest chamber at all once incubation begins. Among many monogamous birds, the task of incubation is shared to a greater or lesser extent between the pair, but for owls the roles are completely separate. Observations of male owls sitting on eggs exist but are very few and far between, and relate to short spells only.

This places great strain on both parents. The male must hunt twice as much as usual, with half of his catches delivered to his mate at the nest. She has a special 'begging call', usually similar in tone to that of a well-grown chick, which she uses to let him know that she requires food. The female inevitably loses fitness and muscle tone from her long spell of inactivity, and must contend with an increasingly oppressive environment as droppings and pellets build up inside the nest. Both roles are vital – should one parent die before the eggs hatch, there is no chance the other could cope alone.

To aid incubation, female owls develop a brood patch, losing feathers from an area of skin on the belly.

The loss of these insulating feathers means that the owl's body heat is more efficiently passed on to the eggs. Despite this, and the generally insulating properties of the nest itself, climatic conditions outside the nest can still have an influence on the temperature the eggs reach and the speed at which the embryo develops. In northern populations of the same species, incubation periods can be several days longer than further south.

Development and hatching Inside each egg, the owl embryo grows and develops rapidly, fuelled by its generous supply of protein-rich egg yolk, with water supplied by the albumen. By the time an egg is laid, the original fertilised egg cell has already divided many times and some differentiated tissue types are forming. After just three days of incubation it has a beating heart, limb buds and the beginnings of eyes, bill, brain and gut. As the chick grows, the yolk sac (attached to the embryo's belly) shrinks, as does the albumen, and the chick takes up more and more space. Eventually, what's left of the yolk sac is drawn into the chick's body.

As the incubation period nears completion, the egg is entirely filled with the chick's body, apart from the air cell, a pocket of air that grows larger during incubation, at one end. About a week before hatching, the chick turns so its head is angled towards the air-cell membrane, and as it continues to grow its head is forced into a tucked position. A day or two before hatching, it straightens its neck and ruptures the air cell with its bill.

At about two weeks old, a Great Horned Owl chick is already showing interest in the outside world.

Asynchronous hatching

The majority of bird species breed once a year, and time it so that the period when there are chicks in the nest coincides with when natural food is most plentiful. This is particularly significant to species that nest in the far north, where the times of plenty are very brief. Birds can only lay one egg at a time, usually one a day or one every two days, but they can ensure that all of them hatch at the same time by only beginning incubation when the clutch is complete. Embryo development is suspended until incubation starts, so the first eggs can be safely left for several days without any problems.

Owls, along with some diurnal raptors, take a different approach, because for them food availability is relatively unpredictable. They begin to incubate their clutch as soon as the first or sometimes second egg is laid. This gives the 'first-born' eggs a head start, and in a large clutch there could be an age difference of a week or more between the oldest and youngest chicks. This will considerably extend the total period that the chicks are in the nest, so what are the benefits?

For the youngest members of the family, there are no benefits at all. If food supplies fall short they will be the ones to suffer – their larger siblings will easily outcompete them and perhaps even kill and eat them. However, the oldest one or two chicks will prosper and grow up strong even if there is a serious food-supply problem – albeit at the expense of the younger chicks. With a brood of same-aged chicks in similar circumstances, all would suffer equally and possibly none would survive or develop into a strong fledgling.

It then begins to breathe for the first time.

When it is ready to hatch, the owl chick begins to peck at the inside of the shell. Its bill is topped with an egg tooth, a small calcified projection that can break the shell. The breakage is a very slow process, allowing the chick to gradually get used to breathing outside air as it works. It gradually chips away a circle of shell, and by bracing its body can lever the two pieces of eggshell apart and hatch. The egg tooth disappears gradually over the first few days after hatching.

CHICK LIFE

Newly hatched owl chicks are feeble, unable to stand or move much. They look curiously unowl-like, as without the facial feathering their bills seem to project forward more. They have a coating of thin down, which helps to insulate them once it has dried off. Their eyes are closed and will remain so for a week or more, before opening gradually over a few days. For the first week or two they need to be kept warm, so their mother continues to sit on them as she did with the eggs – this is now called brooding. Because incubation began with the first egg, the first chick will be alone in the nest for a day or two before the second hatches, and so on.

Feeding The male owl will be aware of the eggs' hatching by the chicks' begging calls, and will hunt more intensively from this point. He brings prey to the nest entrance and passes it to the female who takes it into the nest chamber. Here she prepares it for the chicks to eat by ripping off tiny pieces of meat, which she offers to whichever of the chicks is begging most intensely. It takes the morsel from her bill. The larger chicks naturally receive the first helpings from every meal, and only when they are sated and their calls and begging movements have slowed down will their smaller and younger siblings get their turn. Studies on Barn Owl chicks have shown that hungry small nestlings will not even expend energy in begging until their larger and louder siblings have been fed.

As the owlets grow, the food pieces they are given become larger and start to include roughage of various kinds – fur, feather and bone. Once they are eating this material, usually from the age of a week or two, they will begin to produce pellets. Soon, they will be offered whole small prey items, to be swallowed in one go as adult owls do. This is not without risk, though – owl chicks have been found dead with overlarge prey items stuck in their throats.

Below left: Snowy Owls nest in the open, so the chicks can move around quite freely from an early age.

Below right: A Tawny Owl broods a young chick. As they grow, owlets have less need for parental warmth.

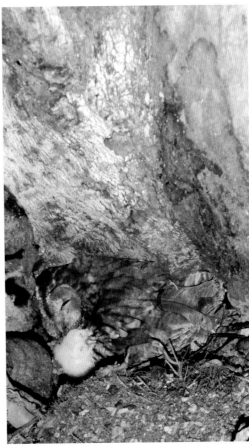

Defending the nest

As discussed in chapter 5, one possible reason for allocating the stay-at-home role to the female is that she is usually the larger of the two, and so better equipped to defend the nest against predators. Owls have several tactics at their disposal to deter predators, some of them cunning, others overtly ferocious.

For small hole-nesting owls, the main threat comes from climbing mustelid species (members of the weasel family), primarily the martens. These animals are adept at scaling tree trunks and have long slim bodies, enabling them to enter quite small tree-holes. The sound of scratching on the trunk of the nest tree has an electrifying effect on a female owl. She moves quickly to the hole entrance to look for the threat. If she sees a marten she may try to camouflage the hole by blocking it with her body. If this fails, she will leave the nest and mob the approaching predator by flying at it, striking and making noisy alarm calls. The male, if he is nearby, will join in.

Owls that nest on the ground in the open may try a pre-emptive distraction display to prevent an approaching predator from discovering the nest. Spotting an approaching human, a female Marsh Owl will leave her nest discreetly, and then fly up in view of the would-be predator, before crashing back down to earth as if injured. Continuing to feign injury, the owl will jump and run away from the nest, trying to draw the predator into pursuit, and thus lead it away. Once the predator is well away from the nest the owl will stage a miraculous 'recovery' and fly away.

Larger owls are particularly vigorous and fearless in nest defence. Both parents will attack intruders with swoops and strikes to the head. Individual birds vary in the intensity of their attacks but are capable of causing serious injury. Researchers ringing Ural Owl nestlings must work in teams, with at least one person ready with a long stick to fend off the parents while the ringers climb up to the nest. Not for nothing do the Swedish call this bird the 'Slaguggla' or 'strike owl'. Even visitors who are leaving the nesting area are at risk of an attack from behind, strong enough to knock them down.

Growing owlets are not defenceless either. From as early as two weeks old they will adopt a defensive posture when threatened, with the wings partly open to make themselves look larger, and if attacked directly will turn onto their backs, ready to kick out with their talons.

With no predator likely to challenge it, the Eurasian Eagle Owl often nests on the ground.

GROWING UP, BRANCHING OUT

In their second or third week, owlets grow a thicker coat of down, called the mesoptile (the owlets themselves may also be called mesoptiles at this age). This is a warmer covering, sufficient to keep the chicks warm without the need for constant brooding by the female. This frees her to leave the nest and go hunting, too, and from now on both parents will hunt to keep the chicks fed. The workload increases as the chicks grow, and it is at this time that you may see normally nocturnal owls out hunting in the daytime. From this point, both parents will directly pass food to the chicks.

Good and bad times With luck, the owl pair will be able to find enough food to sustain their entire brood. However, it is very common for one or two of the youngest chicks to succumb at this time. The age differences between the chicks mean that the oldest will be much stronger, more mobile and louder-voiced than the youngest so will always win the competition for food. If there is a real food shortage, the older ones will also be strong enough to kill their smaller siblings. For the parents, the youngest chicks in a large brood are a bonus if they live, and an insurance policy of extra food for the older chicks when times are hard.

However, there is no automatic antagonism among broods of owlets, as there is with some raptor species (such as the *Aquila* eagles, among which the older chick almost always kills the younger). As long as food is not a serious issue, the owlets get on well and do not bully each other. They may even indulge in mutual preening as they grow older, which helps get rid of loosening down as the true feathers grow up beneath.

If there is plentiful prey, the male does not sit back on his laurels but takes full advantage and kills more than his family needs. He stores the extra prey, sometimes away from the nest but sometimes piled up in its entrance, to be eaten as and when it is needed – but some may go to waste. This, along with the build-up of droppings and pellets inside the nest, makes the chamber quite unpleasant from our perspective, although the owls seem not to have a problem with it. Sometimes debris may be discarded from the hole but only if it gets in the way of efficient nest operation.

The screech owls have a novel way of tackling infestations of flies and other small animals attracted by waste inside the nest. They bring live slender blind-snakes (genus *Leptotyphlops*) to the nest and let them loose inside – the snakes eat the flies and are usually not eaten by the owls in return.

Leaving the nest From about three weeks old in the smaller species, four weeks in the larger ones, the owlets are quite active, able to stand and walk, and may start to venture away from the nest. For species that nest on the ground, such as Short-eared Owl, wandering from the nest happens quite early in the owlets' lives and has an obvious survival benefit – it is easier for one

Above left: Several owl species nest in the safety of large cacti, including the Great Horned Owl.

Above right: A small bird even when fully grown, a young Collared Scops Owl is particularly vulnerable to predation.

Far left: A Long-eared Owl fledgling assumes a tall, elongated posture when alert.

mobile owlet to hide effectively than for a nest of chicks to remain hidden. The chicks disperse into the nearby long grass and lurk quietly in their hiding places, only giving themselves away by calling when a parent arrives with food. Chicks on cliff-face nests may also wander from the nest hollow at an early age.

At this stage, the young owls have developed a loud and distinctive begging call. Hearing this call is often the first clue for biologists looking for evidence of successful breeding, as owls are generally very discreet during the nesting stage of the breeding season.

For tree nesters, leaving the nest may be a little later, but usually occurs when the chicks are still mostly clad in their mesoptile down and unable to fly. They climb from the nest into surrounding branches and there they sit and wait for food deliveries, and sometimes clamber to a new spot. Falling out of the tree is a common hazard for these 'branchers' but is not a disaster. Their wing feathers are quite well grown by this time, enough that flapping should help slow their fall, and there will usually be a soft landing for them on the forest floor. The real danger is that a ground predator like a fox will find them there, but the owlets are strong climbers and can get off the ground into the branches of a nearby bush with ease. Their foot strength will even enable them to climb up tree trunks, beating their wings to help provide lift.

ATTAINING INDEPENDENCE

In most bird species, fledging is synonymous with leaving the nest, as this doesn't happen until the chicks can fly. As we have seen, owls usually leave the nest some time before the first flight, but flying for the first time is the true meaning of fledging. Learning to fly is a natural consequence of young owls using their growing wings while leaping from branch to branch. They are very strong on their feet and as they jump and scramble around they naturally flap for balance. Soon they make longer jumps powered by wingbeats and longer drops slowed down with wings spread for a glide.

Hunting As they learn to fly, so they must learn to hunt. Owlets are naturally curious and attentive to small moving objects. In captivity, they will play like kittens with all sorts of objects, chasing, pouncing and manipulating the 'prey' with feet and bills. In the wild, they will chase and pounce on real prey as well as convincing substitutes like blowing leaves. The first prey taken is usually something slow and easy to catch, such as a beetle or earthworm, and prey like this is likely to form a significant part of the diet even for species that will go on to take mostly vertebrate prey.

Learning to hunt means developing and bringing together a diverse range of skills. Eye–foot coordination is all that is needed to pick a slow beetle from a branch,

Plumage and moult

The first full set of feathers on a bird are referred to as juvenile plumage, and in many birds this plumage is very different to adult plumage, but this is not the case with owls. Once owlets are fully feathered, they usually closely resemble their parents, but there are often subtle but clear differences between juvenile and adult plumage. Juveniles tend to have barred rather than streaked markings on the underparts, and their patterning is often softer and more diffuse than that of the parents.

All owls moult once a year, after the breeding season. All body feathers are replaced in this annual moult, but some flight feathers may be retained, for up to four years. Moult in juvenile owls begins soon after fledging and takes up to three months to complete. The body plumage is replaced, but the long tail and flight feathers are retained. It is therefore possible to tell young from adult owls in their first winter if you have the bird in the hand, by close examination of the pattern and shape of the retained juvenile wing and tail feathers.

SPOTTED OWL

but to take larger prey the young owl needs to develop excellent control in flight and to filter and interpret all of the information that its remarkable ears can pick up. At this stage of life, owls with any defects in these departments will fail to become successful hunters and will not survive.

There is a considerable period of overlap between a young owl making its first successful captures and the cessation of feeding by its parents. The adults may continue to feed fledged chicks for the same amount of time that the youngsters were actually in the nest, or longer. However, the frequency of feeding gradually declines, becoming sporadic, and the young birds must quickly become competent hunters if they are to survive the next phase of their lives.

MOVING OUT

Once the owlets are feeding themselves well, they leave their parents' territory. This is in the interests of both the youngsters and the adults – the young birds need to find a territory of their own, and the adults will need all their home resources for themselves in the coming winter. Sometimes the owlets will move on of their own accord, but they may be driven out by the adults. In all cases they will disperse separately.

Now begins a very difficult time for the young birds, as they go out into unfamiliar territory with their undeveloped hunting skills. The chances are that all the 'best' patches around will already be occupied, so they will have to eke out a living on the fringes. Many will move considerable distances from home to find a patch, even

within species that are extremely sedentary as adults. In species that do habitually make cold-weather movements, it is young birds that move the furthest, as older and more experienced owls will occupy and defend the nearest acceptable feeding grounds to the breeding areas.

Mortality among owls in their first winter is very high. Without a good territory, hunting is difficult, and the owl may also get involved in conflicts with other owls. It may have to hunt through unaccustomed hours (daylight in most species) when it should be roosting, making it more vulnerable to predators. However, opportunities to move into good territories will come up from time to time, and even a territory that is no good for breeding because it has no suitable nest-sites may still have enough food resources to easily sustain a single owl. The more time an owl spends in a particular territory, the better its chances of successfully hunting and surviving there, as it learns where the best hunting spots are, where prey congregates and where is safe to roost. From here it can get familiar with the neighbouring owls, and be ready to make its move if it has an opportunity to replace a neighbour that has died or usurp a weakening rival.

Most young owls reach sexual maturity when they are a year old, and could theoretically breed in their second year of life. However, it is common for breeding to occur for the first time when the owl is two years old (i.e. in its third year of life) as it usually takes this long to find a good enough territory and mate, especially among the larger and longer-lived species.

This Barred Owl is
approaching full adult
plumage, but retains
some down on its head.

Threats and conservation

An individual owl may live for many years. Even the smaller species can easily reach double figures, while larger owls may survive 20 years or more. Captive owls regularly attain this kind of longevity, but in the wild such long-lived birds are very few and far between, with the average lifespan much shorter. The dangers facing individual owls are myriad, but the same goes for all wild living things – as long as the reproductive rate is maintained at a high enough level, the population overall will be stable. Dangers that threaten whole owl species are different and much more serious.

MEASURING VULNERABILITY

Making a meaningful assessment of whether a species is at risk of extinction is an important but difficult task. Only through patient fieldwork carried out year after year can we gain an idea of population size and population trends. Owls present particular challenges because most species are nocturnal with rather secretive ways, and many live in habitats that are difficult to explore and monitor. However, this kind of assessment is necessary to properly prioritise the allocation of resources when planning conservation work.

The Red List The International Union for the Conservation of Nature and Natural Resources (IUCN) has a system for assessing the vulnerability of species to extinction, and assigns a conservation category to each species once it has been evaluated on a global scale. The IUCN Red List is a listing of all species that have been evaluated and are considered to be at risk. Species not yet evaluated or insufficiently evaluated are classed as Data Deficient, meaning that not enough is known yet about their population and circumstances to make a meaningful categorisation.

Species that have been evaluated but are not currently on the Red List are classed as Lower Risk and are categorised into one of four groups. Those considered 'Least Concern' are at the lowest risk of all. Species that meet some but not all of the criteria for inclusion on the Red List may be classed as Near Threatened, while those that are only kept out of the Red List by specific ongoing conservation efforts are classed as Conservation Dependent.

Species are placed on the Red List based on assessments of their population size, geographic range, rate of decline and the nature of the threats they face. Within the Red List there are three categories. Species evaluated as Critically Endangered are those considered to be facing an extremely high risk of extinction in the wild. In the Endangered Category are those considered to face a very high risk of extinction in the wild, while species classed as Vulnerable face a high risk of extinction in the wild. The exact criteria for categorisation are complex, but they can be seen at *www.iucnredlist.org*

Among the 200 to 250 or so described species of owls, six species are evaluated as Critically Endan-

The Boreal Owl is a very widespread species, in no immediate danger from a conservation viewpoint.

Left: Owls of the far north, like the Eurasian Pygmy Owl, may have to travel long distances in cold winters.

gered at the time of writing, and eight as Endangered. A further 16 are on the Vulnerable list, and 22 are Near Threatened. Four are Data Deficient and the remainder are Least Concern.

In addition to the Red List, many countries use their own measures of conservation concern and may use these to justify local conservation efforts, irrespective of the Red List status of the species concerned.

THREATS

It is not common for a species to get into serious trouble because of a single isolated cause. There is often one key problem with a range of interacting factors involved – obviously once a population size falls to a low level, even minor problems can become significant threats. Species restricted to small isolated islands represent special cases. They will have low populations already because of the restricted space available to them and are often very specialised because of limited resources, so any change is likely to have a major impact.

Predation All owls are predators, but few owls are top predators with nothing to fear from other predatory animals. Even the largest species may occasionally be hunted – there is at least one record of a Blakiston's Fish Owl (one of the largest owls in the world) being caught by a Eurasian Lynx, which ambushed the owl while it was hunting along a riverbank. Small owls have numerous potential predators, from mammals such as martens and foxes to other birds of prey including larger owl species.

Although predators account for many owl deaths, they are very rarely a cause of large-scale declines in owl populations. If one prey species becomes scarce, the predator will hunt something that is easier to find instead, and if all prey becomes scarce, the predator population will fall sharply as many will starve. If a predator depends very heavily on just one or a few prey species, its population is at the mercy of its prey. This can be seen with northern owls and the voles and lemmings they eat – the owl populations fall sharply when vole numbers are low.

This system works where a predator and prey species evolve together. Over the generations, they shape each other's evolutionary path, the prey becoming better at avoiding the predator, the predator getting

Forest management methods can make a huge difference to the fortunes of Great Grey Owls.

Enemies in the family

In chapter 3 we looked at the phenomenon of intraguild predation – when species that have a similar ecological niche hunt each other. This is a special case within predator–prey dynamics, as the predator is hunting not just to feed itself but also to remove potential competition. With this double motivation, owls preying on other owls is commoner than would be expected if every case was a chance encounter – there is evidence that owls specifically seek out and kill other owls.

This behaviour drives niche separation between owl species, and sometimes one species may be entirely excluded from certain areas by another. Long-eared Owls in Great Britain occupy a narrower range of habitats than in Ireland, because in Great Britain they are excluded from many areas by the larger and more powerful Tawny Owl, a species that does not occur in Ireland. However, other factors can complicate things and lead to unexpected results – a study in Finland found that Tawny Owls could nest successfully within a short distance of nesting Eurasian Eagle Owls, even though the latter will prey on the former. The abundance of 'normal' prey is certainly a factor in the intensity of intraguild predation, with good prey supplies leading to lower levels of predation.

Intraguild predation involving species that evolved in the same areas is no more likely to threaten a species than 'ordinary' predation, but if two owls that were formerly separated by geographical distance come into contact, either through direct introduction or by habitat change that allows one species to expand its range, there is the potential for conflict.

LONG-EARED OWL

better at catching the prey. In each generation, the predator removes the less well-adapted individuals of the prey species, and, in a roundabout way, the reverse applies too – the less well-adapted predators cannot catch enough prey and so starve to death.

The potential does exist for 'normal' predation to become a problem for certain owl species, when there are other factors involved. In species that have become extremely rare for unrelated reasons, predation can become a significant additional threat, because when a population is down to a few hundred individuals, every bird represents an important part of the remaining population. Habitat change could also make previously successful predator avoidance tactics ineffective and unbalance a once stable relationship between a predator and its prey.

Invasive and introduced species Where things can go badly wrong with predator–prey dynamics is when there is rapid environmental change of some kind, with the result that predatory animals and potential prey that have not had a long evolutionary relationship are suddenly thrown together. There are myriad examples of

introduced predators devastating island ecosystems and wiping out entire species that have not evolved defences against them.

Not all introduced species cause harm by directly preying upon the natives. On Christmas Island, the accidentally introduced Yellow Crazy Ant has established itself very successfully and has caused considerable changes to the island's ecology, by preying upon some invertebrate species, and by protecting and 'farming' leaf-sucking insects, causing die-off of the forest canopy. These changes have impacted upon many native Christmas Island species, including the Christmas Island Hawk Owl *Ninox natalis*, classed as Vulnerable.

Island species are particularly vulnerable to destruction when non-native species arrive, almost invariably in company with humans. The Seychelles Scops Owl *Otus insularis*, evaluated as Endangered, has suffered serious declines due in part to predation from introduced cats and rats, while introduced cats and stoats are implicated in the extinction of the Laughing Owl *Sceloglaux albifacies* in New Zealand. These non-native predators have sometimes been introduced deliber-

Above left: Road collisions are a major cause of mortality in some owl species, including Barn Owl.

Above right: Forest clearance is a potential hazard for many woodland owl species, such as the Ural Owl.

cycle into a shorter northerly summer. The most northerly species of all, the Snowy Owl, could run out of habitat altogether.

Severe winter weather is a problem for many northerly owl species. Some suffer very high losses (mainly of young birds born the previous summer) when prolonged snow and ice make hunting difficult. However, subsequent mild winters will enable them to bounce back. Cyclical fluctuations in rodent numbers have a similar effect, and the northern owls have lived with both of these variables for millennia. Climate change, though, could have severe and lasting effects on weather patterns and rodent breeding cycles, which would have far-reaching consequences for the northern owls.

Direct destruction In some areas, people deliberately kill owls in large numbers, for a variety of reasons. Predatory birds and mammals, especially larger species, have long been viewed as competitors for the kinds of wild animals that we like to eat, so there is a clear motivation to eliminate them, and only relatively recently has a more tolerant attitude begun to prevail in some countries. In Britain, up until the early 20th century gamekeepers routinely shot or trapped every owl on their employer's land, along with other birds of prey, causing very severe declines and (in the case of some raptors) extirpation from the country. Strict protective legislature was introduced during the 20th century, but some illegal persecution continues today.

Removing top predators from an ecosystem can produce unpredictable knock-on effects. Prey species may proliferate to problematic levels, as has happened with deer in Britain after the eradication of large predatory mammals. In assemblages of owl species, removing top species (such as many of the *Bubo* species) may allow the mid-level or meso-predators, species formerly controlled by the top predators, to increase and damage populations of the smallest owl species.

In some cultures, owls are the subject of strong superstitious belief. This may work for and against them, depending on the exact nature of the beliefs. Owls may be protected in recognition of their good work in controlling rodent numbers or reviled and destroyed because of their associations with darkness and their sometimes unearthly calls. In the Mediterranean there is a strong culture of killing large and dramatic birds as trophies, which continues in some areas today, despite legal protection for the birds.

Accidental killing of owls is a serious problem for some species. Owls are often killed in collisions with vehicles, partly because they are active at night and so less easily avoided, and partly because they find roadsides good hunting grounds. Barn and Tawny Owls' habit of flying low along hedgerows places them in the line of danger. And hungry owls that feed on carrion at the roadside are at great risk of becoming road kill themselves. Sometimes owls become caught in traps set for other species, and they are also susceptible to drowning in cattle troughs and other artificial steep-sided water containers.

CONSERVATION

Only with a full understanding of the threats facing a species can we begin to devise ways to tackle the problem – or problems. Some issues are easier to manage than others – often there are conflicting needs or demands to negotiate. However, conservation measures can be very successful if planned and executed well.

Habitat protection and improvement Bringing a halt to ongoing destruction is the first step, and this may require legal changes and perhaps support for local communities that may need to look elsewhere for farmland or firewood. This is far from easy to accomplish, and there are all kinds of political sensitivities that may throw a spanner in the works. Replacing lost habitat is also a high priority where possible. Initiatives such as BirdLife International's Forests of Hope project are working to manage and restore threatened forests in areas of high biodiversity. In Europe, incentives to farmers to manage their land in wildlife-friendly ways will benefit open-country owls.

But conserving entire habitats, and therefore the ecosystems they support, is the surest way to protect the individual species that use them.

Targeted intervention For owls, one very useful form of habitat enhancement is to provide nestboxes. Because suitable natural cavities can be rare, especially where selective woodland management removes old and damaged trees, owl populations may be limited by a lack of nest-sites. One Critically Endangered species that should benefit from nestbox provision is the Anjouan Scops Owl *Otus capnodes* of the Comoros. On Hokkaido island in Japan, most of the remaining pairs of Blakiston's Fish Owls (classed as Endangered) now use nestboxes and also benefit from supplementary feeding.

Other conservation measures that have been used on species in desperate need include translocation, when part or all of a threatened population is moved to an ecologically similar but safer alternative location, and captive breeding, whereby some or all of the population is brought into care and encouraged to breed in captivity. Captive breeding is a plan under consideration to help the Anjouan Scops Owl, giving it a much-needed boost in numbers and a 'pool' of new birds to be released into the wild in the future.

Reintroduction Returning a species to an area from where it has disappeared is known as reintroduction, and can form an important part of a conservation programme. Releasing captive-bred birds into the wild can be spectacularly successful or a total failure. If the factors that caused the species disappearance have not yet been addressed, the newly released birds will in all probability suffer the same fate as their wild-born predecessors. Numerous unofficial releases of Barn Owls in England in the 1980s and 1990s made no difference to the declining wild population.

If the threats have been reduced and eliminated, however, success is much more probable. Because owls will often readily breed in captivity, and can cope well with gradual 'soft' releases into the wild (with food and shelter available for as long as they are required), owl reintroductions often work well. Examples include Ural Owls, which have been reintroduced to Germany, and various projects with Burrowing Owls in North America.

Northern Hawk Owls are well adapted to cope with bad weather, but global climate change could have unpredictable consequences.

Owls and people

Few animals are as familiar and yet as mysterious to us as owls. Books and natural history television programmes help reveal to us the secret world of the owl, and we can see them in person at zoos and falconry displays. However, for most of us, real-life encounters with wild owls remain at least as rare and fleeting as they were before books, television and zoos ever existed. It is easy to see how such a wealth of myth, folklore and superstition came to surround these ghostly birds of the night.

OWL MYTH AND LEGEND

With owls distributed so widely across the world, most countries have their own owl-related mythology. In some cultures owls were worshipped as gods or the associates of gods, and the idea of an owl as a harbinger of death or bad luck is a very common theme. The nature of the mythology surrounding owls in different areas would determine whether people persecuted or protected them.

Gods and god sidekicks The Greek goddess of wisdom (and justice, strength, crafts and several other attributes), Athene or Pallas Athene, gave her name to the Little Owl genus. Her association with owls is apparent through the many depictions of her in which an owl is perched on her head. Athene may be an evolved form of the rather more sinister Lilith from Mesopotamian myth who was a winged and eagle-footed goddess of death, accompanied by twin owls.

In the New World, there were a number of owl or owl-like gods, including Chalchiuhtecolotl of the Aztecs, and the Mayan god of death, Ah-Puch, who wore an owl mask as one of his various disguises. For the Hopi Indians, the Burrowing Owl was the living manifestation of their deity Masau'u, god of night and keeper of game,

This ancient coin from Athens, circa 566 BC, shows a representation of an owl species, probably Little Owl.

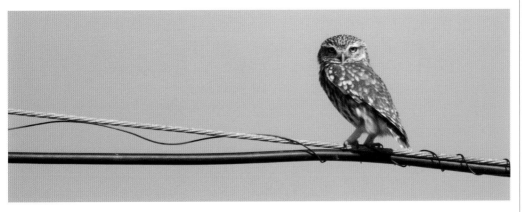

Associated with human habitations since ancient times, the Little Owl features in many folk tales.

Far left: There are many ways that people can help owls. A well-made nestbox makes a safe home for nesting Boreal Owls.

Above: Owls play a key role in J K Rowling's hugely popular world of wizardry.

Above right: The Little Grey Rabbit stories by Alison Uttley feature a typical 'wise owl'.

magicians, to be a wife for the god Lleu. However, she took a lover and the two plotted to murder Lleu, who only just escaped. As punishment, the magicians who created her turned Blodeuwedd (now known as the goddess of betrayal) into a Barn Owl. The goddess's name meant 'flower face', reflective of her origins but also descriptive of the countenance of this beautiful owl.

There are also many tales of mortals being turned into owls. The Inuits have a story that the first owl was a young girl who was suddenly transformed into a bird by magic, and in a panic she flew into the wall of her home, flattening her face and bill and becoming an owl. In Poland it was believed that girls who were married when they died became owls, and had to come out only at night because the birds of the day mobbed them out of jealousy for their beauty.

Owl symbolism In several African countries, owls have had close associations with witches and witchcraft. The Nigerian witch and chief Elullo could change into an owl, and in parts of west Africa the owl is a witch's messenger, and its cry foretells death. Owls were also harbingers of death in parts of India, the Caribbean and Aboriginal Australia. They are assigned the same role in many stories from England (including Shakespeare's *Macbeth*) and elsewhere in Europe – in Sicily, the Common Scops Owl sings for three days by the house of a person about to die. In early Christianity, the owl was a symbol of evil. This perception of owls as evil beings persists in some cultures and many owls are still killed today in the name of these beliefs.

Seeing or hearing an owl is not necessarily considered to be bad news. The Maori recognise two different calls from the island's only owl species, the Morepork or Ruru, one indicating death or trouble, the other announcing good news. In parts of Sulawesi, people listen to owl calls before deciding whether to make a trip, believing some calls indicate travel will be safe but others warn against it. In southern India, the hooting of an owl could bring good or bad luck, depending how many hoots are heard.

Using owl body parts in folk medicine or as magical amulets is widespread. In several cultures eating owl eyes was said to convey the ability to see in the dark, while Sioux Indians believed that wearing owl feathers would have the same effect. In Uzbekistan and neighbouring countries, amulets containing Eurasian Eagle Owl feathers helped ward off evil spirits, for both children and livestock. Many owls are still killed in China and Korea to make traditional medicines of various kinds.

while for the Dakota Indians the same species was a protective god for warriors. The Ainu people of Hokkaido, Japan, traditionally worship owls (along with many other wild animals) as gods, and this reverence is helping to motivate conservation efforts for Blakiston's Fish Owl on the island.

Hindu gods use animals as their vehicles or mounts, and the four-armed goddess Lakshmi – associated with wealth, generosity, wisdom and many other positive traits – sometimes rides a Barn Owl. In Mexico, owls were messengers between the gods of the living and the dead.

Transformations Many mythologies include stories of a person (or a god) being changed into an animal as a punishment or a way to avoid a terrible fate. The Chibcha people of Colombia told the story of Huitaca, the self-indulgent goddess wife of Bochica, who was turned into an owl by her husband when her drunkenness offended him. The Celtic goddess Blodeuwedd or Blodeuedd was created from blossoms by a pair of

OWLS IN FICTION

Because owls are so appealing, they often appear as characters in children's verses and stories, both printed and in film. These fictional owls may be entirely anthropomorphic or in other cases just somewhat repurposed but still very owl-like in their behaviour.

Picture books and nursery rhymes Probably the best known of Edward Lear's nonsense rhymes was the tale of *The Owl and the Pussycat*, an unlikely couple who eloped in a 'beautiful pea-green boat'. A popular nursery rhyme of unclear origins sees nothing sinister in the quiet stillness of a roosting owl:

A wise old owl lived in an oak
The more he saw the less he spoke
The less he spoke the more he heard.
Why can't we all be like that wise old bird?

Popular picture books featuring owls include *The Gruffalo*, by Julia Donaldson and Axel Scheffler, published in 1999, in which a mouse outwits a string of would-be predators, including an owl, by threatening them with an attack from the monstrous Gruffalo. The classic *Little Grey Rabbit* stories by Alison Uttley includes the 1935 story of Wise Owl, who loses his nesting tree in a storm and must be found a new home. *The Owl Who Was Afraid of the Dark*, by Jill Tomlinson, is a very popular 1968 picture book telling the tale of Plop the Barn Owl and how he came to enjoy and appreciate the dark nights that used to frighten him.

The books *Winnie-the-Pooh* (published 1926) and *The House at Pooh Corner* (published 1928), by A. A. Milne, tell the tales of a boy called Christopher Robin and his assorted animal toys, brought to life in his imagination. One of the animals is Owl, the self-proclaimed intelligent member of the team, although time and again the stories show him to be less clever than he thinks he is. In the books' original illustrations by E. H. Shepard, Owl appears to be a Tawny Owl, but in the Disney animated version he becomes a Great Horned Owl.

Stories for older children The *Harry Potter* series by J. K. Rowling, begun in 1997, is a publishing phenomenon, achieving unparalleled popularity. Owls play a key role in this epic story of wizards, witches and magic – they are the messengers that carry all magical communication between the characters, and most individual witches and wizards had an owl of their own. Harry Potter's owl, Hedwig, was a Snowy Owl, but several other species are mentioned by name in the books, including Barn, Common Scops and Tawny.

Arctic communities are very familiar with the Snowy Owl and it has its place in local folklore.

Alan Garner's 1967 novel *The Owl Service* draws on the Welsh folk tale of Blodeuwedd (see above) to conjure up a creepy tale whereby a group of teenagers on holiday inadvertently find themselves re-enacting the myth. The American series *Guardians of Ga'Hoole*, by Kathryn Lasky, begun in 2003, casts owls themselves as the key characters in dark adventure stories rich with legend and symbolism. Many North American species are represented, including Barn, Spotted, Elf, Great Grey and Burrowing Owls.

British author Colin Dann wrote a series of eight children's novels, the first published in 1979, about the animals of Farthing Wood. They were an assortment of typical English woodland animals that had to find a new place to live when Farthing Wood was destroyed by developers; they faced many hazards on their journey and challenges when they set up new homes in White Deer Park. The character of Tawny Owl was introduced in the first book, *The Animals of Farthing Wood*, and made sporadic appearances later on. He was a self-important and opinionated but ultimately well-meaning character, but changed species and sex in the animated versions of the stories, becoming a female of an ear-tufted species (presumably a Long-eared Owl).

A Tawny Owl was a main character in Colin Dann's Farthing Wood series.

Owls on film Owls appear commonly in children's cartoons, typically drawn as plump, wide-eyed figures dispensing wisdom from treetops. A classic example is the character of Big Mama from the Disney film *The Fox and the Hound*, who serves as protector for the title characters. Puppet owls seen in children's television include Hoots the Owl from *Sesame Street* and paper cut-out X the Owl who appeared in the long-running *Mister Rogers' Neighborhood*.

Many of the books mentioned above have been adapted for film or television. The seven *Harry Potter* films feature performances from an array of trained real owls, with some owl characters being assigned to a species for the first time. However, computer-generated imagery (CGI) is used for some of the more difficult scenes. CGI was used to make the 2010 film *Legend of the Guardians*, an adaptation of the *Ga'Hoole* stories.

Owls in captivity No species of owl could be considered to be truly domesticated, but some species have been very widely kept in captivity for centuries. Some are used for falconry, some as decorative cage or aviary birds and others are simply kept for companionship. Many owl species will readily pair up and breed in captivity, providing a supply of captive-bred young owls for the bird-keeping market. However, owls have exacting requirements and can suffer greatly in captivity without them. In the US and other countries there are restrictions on keeping owls as pets.

Falconry Using birds of prey to catch game is a long tradition found in many cultures. The diurnal raptors are much more popular for genuine falconry, as their hunting styles and habits make them much better suited to it from the falconer's point of view. Only a few of the eagle owls (*Bubo* species) have been used extensively in traditional falconry. However, owls of various species are very commonly used in commercial falconry displays.

Training owls for falconry requires different methods to training falcons or hawks. Large species that are active by day are more likely to make successful falconry birds. Smaller and nocturnal owls are very vulnerable to attack by crows and raptors when they fly in daylight, and their tactic of hunting by sound rather than sight makes them difficult to train. Owls are also noted for showing great reluctance to give up a kill to their

handler, and even a very tame eagle owl is capable of causing devastating injury to anyone that annoys it.

The charm of owls makes them a big draw in falconry displays. Training them to fly out and back to the fist is relatively straightforward, and gives the audience the chance to see them on the move at close range. In some shows, trained owls will fly low over the spectators' heads, catch a piece of food in mid-air or chase food items on the ground.

Owls as companions If hand-reared with expertise, owls can be completely trusting of humans and delightfully affectionate. Those who have enjoyed such close relationships with owls will speak of the birds' intelligence, playfulness, ability to communicate and gentle nature. Owning an owl also satisfies the human tendency to want to own things that are beautiful, exotic and a little dangerous. The appeal of a pet owl is obvious, and in the wake of the *Harry Potter* phenomenon, countless children longed to have one of their own.

However, owls are not domesticated animals and to live happily in captivity they need a specialised diet, an aviary large enough for them to fly in, and all handling supervised by experienced people who understand that every owl, however tame it appears, is not so far removed from the wild that it can be treated with anything less than total respect. Unfortunately, many would-be owl owners found that they could not meet these requirements, and wildlife rescue centres are full of unfortunate owls that were kept in totally unsuitable conditions. Many more must die from neglect, while others escape from their cages or are dumped in the wild. If this happens, the smaller species especially are highly likely to starve or be taken by larger predators. Larger and non-native owls may fare better, but if they become established in the wild, as may be happening with Bengal (*Bubo bengalensis*) and Eurasian Eagle Owls in Britain, they could unbalance and damage ecosystems.

Rescue centres can provide a good opportunity for people to enjoy close-up encounters with owls, and they also have a role to play in educating the public – both about owls in general and why it is not a good idea to keep them as pets. Most rescued 'pet' owls will have to live in captivity for ever, but rehabilitation and release may sometimes be possible.

Barn Owls are popular in falconry shows because of their beauty and (usually) tractable natures.

Owls of the world

Owls stand out among birds in several obvious ways, but across the whole range of owl species there is rather less variety than is observed in comparable bird orders. The owl blueprint is evidently quite successful with little need for modification from continent to continent, and habitat to habitat. However, every owl species does have its own unique set of traits and attributes. The second half of this book looks in detail, species by species, at owls of the Holarctic regions of North America and Eurasia, and this chapter gives an overview of the rest of the owl world.

THE MAIN GROUPINGS

The owl order Strigiformes is large and diverse, containing some 200 to 250 species (the exact number varying according to taxonomy). Within this order are just two families, one much larger than the other. They are the Tytonidae or barn owls, and the Strigidae or typical owls. Each family can be divided further, into distinct clades between family and genus level. These clades are designated subfamilies, each containing related groups of genera. Tytonidae comprises two subfamilies, Tytoninae and Phodilinae. Each subfamily contains just one genus – *Tyto* in Tytoninae and *Phodilus* in Phodilinae. The much larger and more complex family Strigidae comprises three subfamilies – Surniinae, Ninoxinae and Striginae, and within Striginae in particular there are further distinct groupings of genera.

Owl genera Species are placed in the same genus based on similarities in shape, plumage, voice and behaviour, with the real clincher being DNA analysis that shows the two share a recent common ancestor, but there is no definitive set of rules to determine whether two species belong in the same genus. As a result, species are often shuffled from one genus to another, with new genera created and old ones subsumed in the light of new information yielded by lab studies of DNA. Examples of this within owls include the subsuming of the fish-owl genus *Ketupa* into the eagle-owl genus *Bubo*, and the separation of the screech owls from the scops-owl genus *Otus* into a new genus of their own – *Megascops*. The owl families and genera, and some representative species, are discussed below.

TYTONIDAE AND ITS GENERA

The species within the small family Tytonidae all look rather alike and clearly different to the typical owls. They all have elongated faces, the bill set noticeably further down from the eyes than in the typical owls, and when relaxed their facial disks are broader across the top than the bottom, tapering to a point and looking heart-shaped. There are about 15–17 species of *Tyto*, the sole genus in the subfamily Tytoninae, but just two of *Phodilus* – the bay owls that make up the subfamily Phodilinae.

The Little Owl is the only member of its genus to occur in Europe.

Left: The Eurasian Eagle Owl is a typical example of the large genus *Bubo*.

Barn owls The Barn Owl is the most widespread member of the genus *Tyto*, and all of the other species are very Barn Owl-like in appearance, with elongated heart-shaped faces, relatively small dark eyes, long, mostly bare legs (feathered only at the top of the tarsi) and long wings. They have serrations on the undersides of their middle toe, used to 'comb' their very soft sound-proofed plumage. The *Tyto* owls' pellets are unlike those of the typical owls, having a smooth and shiny silk-like coating.

Some of the most distinctive *Tyto* species include the Madagascar Red Owl *Tyto soumagnei*, a rather small and uniformly orange species found only in northeast Madagascar. The two species of sooty owls (Greater *T. tenebricosa* and Lesser *T. multipunctata*) from Australasia both have dark charcoal-grey plumage – they are forest birds and so have shorter and broader wings than the open-country barn owls. There are several *Tyto* species that are restricted to small islands or island groups in the Pacific, including some classed as Endangered, such as the Lesser Masked Owl *T. sororcula* and Minahassa Masked Owl *T. inexspectata*, and some Vulnerable, like the Golden Masked Owl *T. aurantia* and Manus Masked Owl *T. manusi.*

Bay owls This genus comprises two species of bay owls, the Sri Lanka Bay Owl *Phodilus assimilis* and Oriental Bay Owl *P. badius*. Bay owls are rather small and thickset owls, with large dark eyes, shaggy feathering on the tarsi, and a distinctive face shape whereby the facial disk extends upwards to form two peaks above the eyes, like rudimentary ear-tufts. Both species have barn owl-like dappled orange and golden plumage tones. Other *Tyto*-like traits include a serrated middle claw and pellets with a silky coating.

NINOXINAE

This is the smallest subfamily of typical owls, containing the genera *Ninox* and probably also *Uroglaux*. They form a very distinctive group. There are some 26 species, all but one restricted to Australasia. They are usually known as the hawk owls, though care should be taken to avoid confusion with the unrelated Northern Hawk Owl.

Hawk owls A typical *Ninox* owl is slim and long-tailed, rather small-headed, and has a distinctive face with no clear definition to its facial disk, but very prominent eyes and bill. From the front, the cere is visible – this is a pad of skin at the upper bill base, present on all owls but partly or completely concealed by feathers in many – and the front-set nostrils can be seen within the cere (in other owls the nostrils are at the cere sides and are not as noticeable in a head-on view). In profile, the bill projects forward clear of the facial plumage, and the entire face is more pointed than that of other owls. The eyes look rather protuberant, unlike the often deepset and heavy-lidded eyes of other owls, and are usually yellow.

Most of the owls found on the Australian continent are *Ninox* species. They include the widespread Southern Boobook *Ninox boobook*, a common bird that utters its cuckoo-like two-note song on clear nights. Its equivalent in New Zealand is the Morepork, often known by its Maori name of Ruru *N. novaeseelandiae* – both names describe the bird's song. The very large and hawk-like Powerful Owl *N. strenua* is Australia's largest species.

The only non-Australasian *Ninox* owl is the Madagascar Hawk Owl *N. superciliaris*, found in the northeast and south-west of Madagascar. It is the only dark-eyed *Ninox*. Many of the hawk owls are island endemics, and some are classed as Endangered or Vulnerable. The Papuan Hawk Owl *Uroglaux dimorpha* is the sole member of its genus, and while its relationships are not well studied it appears to be very similar to the *Ninox* hawk owls.

The Oriental Bay Owl's unusual face shape distinguishes it from the barn owls.

SURNIINAE

The subfamily Surniinae contains the little owls (*Athene*), the forest owls (*Aegolius*), the pygmy owls (*Glaucidium*), the Northern Hawk Owl *Surnia ulula* and a few other small genera. Although most of the Surniinae species are small birds, they are all powerful predators for their size. They lack ear-tufts, but some have 'false eye' markings on the backs of their heads.

Little owls This small group of species belongs to the genus *Athene* and includes the Little Owl of Eurasia and the Burrowing Owl of America. These are small and dumpy but long-legged owls, with fierce expressions. Some species are known as 'owlets', such as the Forest Owlet of India, which was thought to be extinct, known only from museum skins, until it was refound in 1997. Its current status is uncertain but it seems to be extremely rare with a tiny geographic range. The Spotted Owlet *A. brama* is also found in India but is much more widespread, replacing the Little Owl in southern and south-eastern Asia.

Pygmy owls The large genus *Glaucidium* contains the pygmy owls, which are small or very small owls with quite long tails that they often cock and flick. They have rather small yellow eyes and poorly marked facial disks, and have false-eye markings. They are fearsome predators, punching well above their weight in terms of prey size, and attract particularly intense mobbing from other birds when found at their roosts. As with the *Athene* owls, some species are known as 'owlets'. The Eurasian and Northern Pygmy Owls have very extensive ranges across Eurasia and North America respectively, but there are more than 20 other species, mainly in the New World. Many of them have small ranges and are separated by regions of unsuitable habitat.

One of the smallest owls in the world is the Tamaulipas Pygmy Owl *Glaucidium sanchezi*, which has a tiny range in Mexico. It is the size of the Elf Owl but slightly heavier. The Red-chested Owlet *G. tephronotum* is one of the most colourful owls with its bright rufous breast-sides. It is found in Africa, along with the more wide-spread Pearl-spotted Owlet *G. perlatum*, a species noted for the very dramatic mobbing reaction it attracts from other birds. A good imitation of the call of a Pearl-spotted Owlet will conjure dozens of agitated small birds out of the bush.

Forest owls Species that make up the small genus *Aegolius* are known as forest owls, and include the saw-whet owls of America and the Boreal or Teng-malm's Owl of the Holarctic. They are small, square-headed owls with penetrating stares, and of the four species the Buff-fronted Owl *Aegolius harrisii* of South America is the most striking with its unmarked cream underparts, whitish facial disk and heavy, arched black-ish 'eyebrows'.

Other Surniinae genera The species in *Taenioglaux* were formerly classified in *Glaucidium*, but all show some consistent differences that warrant their separation into a new genus. They lack false eyes on the nape, and when agitated swing their tails rather than flick them. They also have different calls. All are Old World species.

Above left: New Zealand's only extant native owl is the Morepork or Ruru.

Above right: The Southern Boobook is one of Australia's best-known owls.

Above left: Widespread south of the Sahara, the African Wood Owl is a typical *Strix* species.

Above right: The African Barred Owlet is a small but notoriously fierce species of Sub-Saharan Africa.

The Long-whiskered Owlet *Xenoglaux loweryi* is a peculiar owl, the only species in its genus. It is a very small owl, only slightly heavier than the Elf Owl, and is confined to a small pocket of forest in the eastern Andes. It gets its name from the long, bristly feathers around its bill and the sides of its head. It is known only from three specimens and a handful of observations. The Elf Owl of southern North America is the smallest owl species and is also the only member of its genus (*Micrathene*). The unusual Northern Hawk Owl is the sole representative of the genus *Surnia*.

STRIGINAE

All the rest of the typical owls belong to the large and diverse subfamily Striginae, including the scops owls (*Otus*), wood owls (*Strix*), eagle owls (*Bubo*), screech owls (*Megascops*), eared owls (*Asio*) and a range of other smaller genera. All of the owls that bear ear-tufts are in the subfamily Striginae.

Wood owls Members of the genus *Strix* are sometimes known as the wood owls. There are 20 or so different species, and they include some widespread and well-known species, though by nature the wood owls are rather shy and elusive. They inhabit woodland of various kinds and range in size from medium to large. None have ear-tufts. There are representatives across most of the world, except for parts of Australasia where they are replaced by the *Ninox* owls.

Some of the *Strix* species of Central and South America are very striking. The Black-and-white Owl *Strix nigrolineata* has black upperparts and white, black-barred underparts, a black facial disk and contrasting bright yellow bill. The Black-banded Owl *Strix huhula* is entirely blackish with delicate white barring. These two species, together with the Mottled Owl *S. virgata* and Rufous-banded Owl *S. albitarsis*, are sometimes split from *Strix* into a new genus, *Ciccaba*.

Genera related to Strix Genetic study places three other genera in a clade with the wood owls. The four spectacled owl species of the genus *Pulsatrix* are found in Central and South America. All are strongly marked with dark and light facial patterns. They all have quite extensive forest ranges but only the Spectacled Owl is reasonably well known. With its dark face and whitish 'spectacle' markings, together with a blackish breast-band and pale cream belly, it is one of the most boldly marked of all owls.

The little-known Maned Owl is the only species in the genus *Jubula*. It is found in sub-Saharan west Africa. It is a stocky, rufous owl with a strongly outlined facial disk, and long, droopy ear-tufts that extend into shaggy plumage down the neck-sides. In South America is another unique species, the Crested Owl *Lophostrix cristata*, also alone in its genus. It is a large, dark owl with rather uniform plumage except for its long, mostly white ear-tufts.

Eagle owls and allies The large genus *Bubo* is made up of very large owls, although some of the more atypical forms – the Snowy Owl and the fish owls – were classified in separate genera before studies showed they were too closely allied to the 'core' *Bubo* species to warrant this separation. Most *Bubo* owls are very big and heavy birds (the genus includes the two largest species in the world, the Eurasian Eagle Owl and Blakiston's Fish Owl) with prominent ear-tufts. The Snowy Owl is the outlier in terms of appearance, with white plumage and imperceptibly tiny ear-tufts.

Besides the well-known northern hemisphere *Bubo* species, other notable members of the group include Verreaux's Eagle Owl *Bubo lacteus* of southern Africa, a softly toned grey bird with short ear-tufts, dark eyes, a boldly outlined facial disk and striking pink eyelids. It is the largest and heaviest African owl, and is a mighty predator capable of taking monkeys and baby Warthogs. Also from Africa, Pel's Fishing Owl *B. peli* is, like the other two fishing-owl species, a tuftless, dark-eyed and rather pale bird of riverside forests.

There are several other *Bubo* species found only in Africa, including the rare Usambara Eagle Owl.

Scops owls The scops owls that form the large genus *Otus* are mainly small, slim, short-winged owls with prominent ear-tufts, widely distributed in the Old World. Depending on taxonomy, about 50 different species are recognised, more if the New World species are included rather than being separated into *Megascops* (see below). This genus shows a very high tendency for island endemism, with neighbouring but isolated populations showing subtle but consistent differences that warrant full species status.

No fewer than six different isolated *Otus* species are found on Madagascar and nearby islands. In southeastern Asia there are separate endemic species on numerous islands, including Sulawesi, Pemba, Palawan, Wetar, Sula, Simeulue, Enggano and Kalidupa. Several of these are on the IUCN Red List of conservation concern, as their populations are so small. A number of new scops-owl species have also been discovered in relatively recent times, including the Serendib Scops Owl *Otus thilohoffmanni* of Sri Lanka, which was described in 2004, and the Sokoke Scops Owl *Otus ireneae* of east Africa, found in 1966. Recent taxonomic research also supports

With its bold face pattern, the Spectacled Owl of the Americas is unmistakable.

the splitting of the Philippine Scops Owl *Otus megalotis* into three species.

Megascops Formerly classed in *Otus*, the American screech owls certainly bear a resemblance to the scops owls but their genetic profiles indicate that they should be classified separately. There are some 27 species. Notable features they share are similar song structures (with two distinct song forms per species). Most species are larger than the scops owls, some considerably so.

Other scops-like owls The *Otus* and *Megascops* owls are thought to share a lineage with a few other smaller genera, many of which have had considerable taxonomic shuffling over the years. The two white-faced owls (genus *Ptilopsis*) are still often known as the 'white-faced scops owls' although they are no longer classified in *Otus*. Both species occur in Africa. The Northern White-faced Owl *Ptilopsis leucotis* is found in a band from Senegal and Gambia across to northern Kenya, and is replaced south of here by the Southern White-faced Owl *P. granti*. They are striking small grey owls with white faces, heavily outlined with black. They have short ear-tufts and bright orange eyes.

The Giant Scops Owl *Mimizuku gurneyi* is the sole representative of its genus, although there is some evidence that it should be reclassified into *Otus*. Found in the southern Philippines, it is a large, dark-faced owl with strong patterning and large ear-tufts. Two other oddball owls that are currently considered to belong to the *Otus* and *Megascops* clade but are alone in their respective genera are the Cuban Bare-legged Owl *Gymnoglaux lawrencii* and the Palau Scops Owl *Pyrroglaux podarginus*. The Cuban Bare-legged Owl, a bird of rocky countryside on Cuba, looks strikingly similar to the *Athene* owls in general shape, and may prove to be more closely allied to that genus than to *Otus*. The Palau Scops Owl is a little-known species whose taxonomic position is yet to be established. It is a stocky, tuftless owl with finely patterned rufous plumage.

Eared owls and their allies The final taxonomic grouping within Striginae comprises the eared owls of the genus *Asio*, and their relatives. The genus *Asio* is small, with just eight species, but between them those species have a very extensive geographic range, with the Long-eared and Short-eared Owls being particularly widespread. The remaining *Asio* species include the Stygian Owl *Asio stygius*, which looks like a darker and greyer Long-eared Owl and occurs widely in South and Central America, the Marsh Owl of southern Africa and the Striped Owl *Asio clamator*, another widespread South American species.

A couple of other species are likely to be closely related to *Asio* based on appearance, although both require further study. They are the Fearful Owl *Nesasio solomonensis* of Bougainville and the Solomon Islands, which resembles a Short-eared Owl but lacks ear-tufts and has strikingly large and strong talons and bill, and the Jamaican Owl *Pseudoscops grammicus*, a dark-eyed, tufted woodland species, endemic to Jamaica.

The Barking Owl is a secretive but noisy Australian species.

Far right: The Southern White-faced Owl is a distinctive and charming species.

GLOSSARY

Brood (*noun*) All of the chicks of one nest.
(*verb*) To keep young chicks warm with body heat.

Cache To hide food for later consumption. Also used as a noun to describe a store of food.

Call Any vocalisation not defined as a song.

Camouflage Plumage patterning that helps an owl to blend in with the colours and tones of its background, so it is less easily spotted.

Clade Any monophyletic group of species.

Clinal variation Gradual variation in appearance within a species across its range.

Clutch All of the eggs in one nest.

Courtship All interactions between a pair of owls at the start of the breeding season.

Coverts The rows of short rounded feathers that cover the inner part of the wing, becoming longer towards the outer edge of the wing.

Crepuscular Active at dawn and dusk.

Cryptic Describes plumage with a complex camouflaged pattern.

Dispersal Of newly independent young owls, moving away from the birthplace to look for a new territory.

Ear-tufts Tufts of feathers on the crown, one positioned above each eye.

Eyebrows A line of feathers above the eye, in a contrasting colour.

Facial disk The rounded surround of an owl's face, usually outlined with stiff short feathers in a contrasting colour.

Fledge Of young birds, to make the first flight (in owls, this often occurs some time after leaving the nest).

Flight feathers The long wing feathers – primaries and secondaries.

Flush To drive or frighten a hiding animal out of cover.

Gizzard A muscular organ in the digestive tract, within which pellets are formed.

Incubation The process of sitting on a clutch of eggs, thereby maintaining them at a sufficiently high temperature to allow the embryo to survive and develop.

Incubation period The length of time it takes for an incubated egg to hatch.

Irruption An unpredictable mass movement of birds away from their breeding grounds, usually observed in winter in response to severe food shortages.

Mesoptile The second thick downy plumage of a young owl, often still present on the whole body when the owlet leaves the nest.

Migration A regular and predictable annual journey that a species makes to wintering grounds that are separate from where it breeds.

Monophyletic A group of species that descend from one common ancestor. Every genus, subfamily, family, order and so on is monophyletic.

Monotypic A species that has no subspecies variation.

Mustelid A predatory mammal of the family Mustelidae – common examples include weasels, stoats and martens.

Nominate In species with two or more subspecies, the nominate is the subspecies that was the first one to be formally named. A nominate's subspecific name is the same as the species name. For example, in the Northern Saw-whet Owl the nominate subspecies is *Aegolius acadicus acadicus*.

Pellet A compacted mass composed of the indigestible parts of prey the owl has eaten – fur, feathers, bones and the hard parts of insects. The owl regurgitates pellets regularly.

Primaries The outermost set of flight feathers, corresponding to the 'hand' of the wing.

Rodent A small, gnawing mammal of the order Rodentia – common examples include mice, voles, rats, lemmings and squirrels.

Roost To be asleep or inactive – also used as a noun to describe the hiding place where an owl spends its inactive hours.

Scapular line A line of contrasting pale feathers along the scapular feathers, on the sides of the back.

Secondaries The inner set of flight feathers, corresponding to the 'arm' of the wing.

Song A specialised call given to attract a mate or proclaim ownership of a territory.

Subspecies A consistently distinct population within a species, but not yet sufficiently genetically distinct to be considered a separate species. It is denoted with a third part of the scientific name, for example, the Northern Saw-whet Owl has two subspecies: *Aegolius acadicus acadicus* and *Aegolius acadicus brooksi*.

Tapetum A reflective layer behind the retina in an owl's eye, which reflects light back towards the retina, allowing for more acute vision in low light.

Tarsi – singular: tarsus The visible part of the legs below the ankle bends, fully or partly covered with feathers in many owls.

Territory An area of land which an owl or pair of owls defends from others of the same species, and which usually contains the nest-site.

Wing-loading Calculated by comparing the body size and weight to the wing area. Many owls show low wing-loading (large wings relative to body size), enabling them to use a slow flapping flight with less energy expenditure.

The species accounts

Great Grey Owl

The remainder of this book takes a detailed look at each of the owl species that occur across northern Eurasia and North America. These 41 species include some of the best-known owls in the world, but also a fair few that are more obscure. This introduction serves to outline the rationale behind the species selection and explain the structure of the accounts.

THE RANGE COVERED
Biologists divide the land surface of the world into several terrestrial ecozones, which reflect the distribution of plant and animal groups. The assemblage of life within each ecozone is broadly isolated from all other neighbouring ecozones by natural boundaries of one kind or another, usually stretches of ocean, mountain ranges or extensive deserts (although there is always some overlap, especially among birds as they are the most mobile of organisms). The Holarctic comprises two ecozones – the Palearctic, which includes Europe, most of Asia (south to the Himalayas) and North Africa, and the Nearctic, which covers North America and Greenland, south to the highlands of Mexico. All the owls that occur within the Holarctic are included, even though some of them barely edge into the region and have the bulk of their range further south.

ACCOUNT LENGTHS
Some owls are very well studied, while others are hardly known at all. Species such as Barn and Short-eared Owl have a very extensive range and have been studied in great detail – although even for these birds there is still a great deal to learn. For others, we have barely scratched the surface. Lesser known species tend to be those with small ranges in inaccessible areas. We can infer certain things about their habits by looking at better-known species that are closely related, but without more fieldwork projects these inferences remain unconfirmed.

The accounts in this book range in length from one to four pages, and their length broadly reflects the information available on each species. The most well-known species are given more space to allow for the inclusion of extra information, mainly on hunting and breeding behaviour. The species are grouped together according to their genera as far as has been possible within the constraints of the book's design.

STRUCTURE OF THE SPECIES ACCOUNTS
Each account follows the same format. The sections are as follows:

Range describes the owl's full distribution, including areas outside the Holarctic.

Evolution and relationships examines the taxonomic position of the species (including its genus), according to current available data.

Description gives a detailed account of the bird's shape, plumage pattern, including any known colour morphs, colour of bare parts, any striking features noticeable only in flight and a description of the juvenile. In polytypic species, the description covers the nominate subspecies.

Geographical variation gives details of some or all recognised subspecies, where they occur, and how they differ from the nominate form. There is much disagreement, however, on subspecies, and taxonomic changes are frequent. As research proceeds, new evidence is constantly appearing that results in some subspecies being promoted to full species status while others are subsumed.

Movements and migration looks at both regular migratory behaviour and unpredictable movements such as dispersal in cold winter weather.

Voice describes songs and calls of adults and chicks.

Habitat gives details of the kinds of terrain the species uses, including seasonal variation where relevant.

Behaviour, hunting and diet looks at the species' habits, roosting behaviour, its preferred hunting tactics and the range of prey it takes, including exceptional prey records.

Breeding provides an overview of the species' reproductive habits, including when breeding behaviour begins, preferred nest-sites, incubation period, when and how the chicks leave the nest and when they fly for the first time.

Status and conservation looks at the bird's population size, past and current population trends, details of any threats it may face and any conservation initiatives in place to benefit it. For each species the current IUCN Red List status is provided.

Although efforts have been taken to avoid excessive jargon, some special terms are used in the species accounts – these are fully explained in the Glossary.

Barn Owl
Tyto alba

Patterned in white, gold and soft grey, the Barn Owl is both distinctive and beautiful.

Right: Hunting birds are often active before sunset, especially in the summer months.

Size 34cm

Range This is a very widespread owl, although some authorities separate it into two or more species. Its range encompasses most of the US states and the extreme south-east and south-west of Canada, down through Central America and most of the Caribbean and through the entire South American mainland. Its Old World distribution extends from most of Europe (except the far north) and north Africa, to parts of the Middle East. There are further large discontinuous populations in the Indian subcontinent and south-east Asia, and in most of sub-Saharan Africa and Madagascar. It is also found on various island groups including the Cape Verde Islands, Canaries and Comoros.

Evolution and relationships As mentioned in the Eastern Grass Owl account (page 102), the genus *Tyto* is part of the family Tytonidae and comprises all of the barn owls, as distinct from the typical owls in the family Strigidae. The Barn Owl is sometimes split into as many as six distinct species, based on DNA and other evidence.

Description This is a medium-sized owl with a long face, large head, long wings and long legs. Its main colours are grey, gingery or sandy yellow and white, although the intensity of colour is highly variable. The upperpart feathers are mainly yellow-based and grey-tipped, with blackish and whitish speckles. The yellow sections form a broad pale band across the secondaries. The nape and crown are finely speckled. The underparts vary from white through yellowish to a deep reddish-ochre tone, and may be completely plain, finely speckled or quite heavily spotted with dark.

The facial disk is outlined with dark speckles and is distinctly heart-shaped, but becomes much rounder when the bird is alert. The long shape of the face makes the eyes and bill look more separated than in the typical owls. The disk is usually unmarked except for a small dark area around the eyes, and paler than the underside. The eyes are black and rather small, the long-looking bill is horn-coloured and the claws are greyish. The legs are thinly feathered at the top of the tarsi but the feathering thins out to nothing halfway down. In flight it looks pale and long-winged, with a prominent pale panel in the secondaries. Juveniles resemble adults.

Geographical variation A large number of subspecies have been recognised, some of which probably warrant elevation to full species status, while others may not be valid but represent individual variation. The nominate is found across western Europe including Britain, and is a warm-toned, pale form. It meets *T. a. guttata*, a darker-breasted form, in central Europe, and the very pale *T. a. ernesti* in the Mediterranean. In North America is the larger *T. a. furcata*, often split as a separate species with a number of subspecies of its own. The same goes for the Australian *T. a. delicatula*, a very pale and grey-toned form. Distinctive island subspecies like the very dark and heavily spotted *T. a. insularis* of the Lesser Antilles and the bright rufous *T. a. deroepstorffi* of the Andamans are also candidates for full species status.

Movements and migration This owl is generally sedentary, not usually ranging into latitudes that become completely inhospitable in winter. However, it will wander in response to prey shortages, sometimes

considerable distances (more than 450km in extreme cases). Birds at the northern and southern limits of the world range are most likely to have cause to wander, but even they will remain on their territories all year if they can. Juveniles may roam widely but tend to settle not far from their place of birth.

Voice The most often described call is a rather blood-curdling harsh shriek. This appears to be the species' territorial song, as it is given at regular intervals by both males and females from different points of the territory, especially at the start of the breeding season. Challenges to the territory from other Barn Owls are met with a high-pitched purring call, while the begging call (given by chicks but also females soliciting food from their mates) makes a peculiar whooshing or snoring sound.

Habitat An open-country species, the Barn Owl is found in a wide range of landscapes. In Britain it is mainly associated with extensive grassy meadows but will also hunt over mixed farmland, grassy road verges, coastal grazing marsh, reedbeds, heathland and moor, and uses equivalent habitats elsewhere. It is often found on woodland edges or places with scattered trees, but avoids closed woodland. It also avoids mountainous areas and very arid regions. It is associated with human habitation in many areas, especially where farming practices are not intensified, buildings provide suitable nest-sites and agricultural activity can attract good numbers of rodents.

Behaviour, hunting and diet The Barn Owl is primarily nocturnal but the hunting period may begin before dusk and continue until after dawn, especially in summer when nights are short and there are fast-growing chicks to provision. In winter, young birds struggling

This species hunts birds from time to time, especially small, ground-dwelling species.

to find enough prey will hunt in broad daylight. Usually, though, it spends the daylight hours roosting in a bush or tree, or around the nest-site.

This species is quite similar ecologically to the *Asio* owls and also to the harriers – all are hunters of small rodents and other ground-dwelling animals of open countryside. Like those species, the Barn Owl hunts mainly in flight, working across a field with a low and slow searching flight with much flapping and hovering. Its large wing area relative to a light body weight gives it a very low wing-loading, which helps make its hunting flight less energetically expensive, and it also flies into the breeze to make use of the uplift. When there is no wind, or too much, it may switch to still-hunting from a perch.

Prey is usually detected by sound, and the owl uses its precision hearing skills to judge an accurate pounce into the long grass. It may also chase bats or birds in flight, and has been seen to fly alongside hedgerows and touch the hedge with its wing-tips to flush out birds hiding inside. Birds make up a small proportion of the diet, with most of it composed of whatever the common local vole and mouse species are – these typically form 60–90 per cent of the diet by weight.

Breeding Barn Owls form lasting pair bonds, both birds staying on territory all year (unless forced away by food shortages). They renew their bond in late winter with the screeching 'song' and also by aerial display, including circular chases. They may preen each other's head and neck plumage before and after copulation.

Nest-sites are used year after year and are jealously protected – good sites are often at a premium, and the Barn Owl may have to fend off challenges not just from other Barn Owls but from different bird species – in western Europe, nest-site disputes between Barn Owls,

Kestrels, Jackdaws and Stock Doves have been observed. The site may be a tree cavity but more often is a ledge inside a disused building, accessed from a high opening. It will readily use nestboxes, and anyone renovating or demolishing an old building that was home to a Barn Owl pair should consider putting up a suitable nestbox as an alternative for the birds.

The clutch is usually between four and seven eggs but may reach double figures in peak years. Eggs are laid every two days, and incubation begins with the first, so there can be substantial age differences within the same brood, and the younger chicks are highly likely to fall victim to their older siblings if food supplies run short. The male hunts for the family alone while the female incubates for about 35 days and broods the chicks for another couple of weeks. Thereafter, both parents hunt for the chicks, which begin to wander around and outside the nest-hole (if its location permits this) by about six weeks old. They can fly from about two months and are more or less independent at the age of three months, whereupon the parents drive them away and in some areas may then begin a second brood.

Status and conservation With such a huge world range, inevitably this species is doing well in some areas and badly in others. It is known to suffer heavy losses in severe winters and to be very badly affected by agricultural intensification. Locally, its status may be of considerable concern – in Britain, for example, it is amber-listed as a species of heightened conservation concern, though it has responded well to various initiatives to improve its habitats and at the time of writing has shown a considerable recovery. However, globally its huge range and population of at least 5 million individuals places it in the Least Concern category.

Above left: Listening and watching, a Barn Owl hangs in the breeze before dropping down to catch prey.

Above right: Small mammal prey is usually swallowed in one go, head first.

Eastern Grass Owl
Tyto longimembris

Size 38–42cm

Range This owl occurs patchily from southern India east and south through China to Australia, and on the Philippines, Sulawesi, New Guinea and some other Pacific islands.

Evolution and relationships The barn owls all belong to the family Tytonidae, which contains the two genera *Tyto* and *Phodilus*. The Eastern Grass Owl shares a lineage with other Australasian species in *Tyto*.

Description This is a typical barn owl, slim, large-headed and long-legged. Its general colour tone varies from whitish to rich reddish, the upperparts strongly patterned with grey bars and mottling, the underparts almost unmarked. The facial disk is heart-shaped and plain, and darker in females than males. The small eyes are dark, the bill and claws horn-coloured, the tarsi feathered on the upper halves only. In flight it looks pale, long-winged and long-headed. Juveniles are like adults.

Geographical variation Between four and seven subspecies are recognised, with individual variation confusing the picture somewhat. *T. l. chinensis* of China and Taiwan can be very rich red-brown on the face and breast. The Philippine form *T. l. amauronota* is pale but has a very grey facial disk.

Movements and migration Most populations are generally sedentary but will wander short or long distances to exploit exceptional prey abundance.

Voice It is not very vocal but produces hissing, snoring and screeching calls during the breeding season.

Habitat It frequents open fields with long grass, both lowland and upland, and in dry and wet areas – as a ground-nester it does not require trees.

Behaviour, hunting and diet It is mainly nocturnal, though may be seen hunting by day (especially early or late) when feeding well-grown chicks or when prey is hard to find. When roosting it sits on the ground among long grass. It hunts on the wing with a slow, low patrolling flight, often hovering (assisted by headwinds) before the final pounce. Prey is mainly small rodents, but opportunistically other small vertebrates and invertebrates.

Breeding The nest is made in a sheltered hollow on the ground. The female lays three to eight eggs and incubates them alone, starting at the first egg. The incubation period is 42 days. The chicks fledge at two months old.

Status and conservation It is declining in some areas, suffering the effects of prey shortages where pesticide use is high. Population size is unknown but it has a wide distribution and is classed as Least Concern.

With a typical long *Tyto* face, the Eastern Grass Owl's close relationship to the Barn Owl is obvious.

Mountain Scops Owl
Otus spilocephalus

Size 18cm

Range This owl occurs in forested uplands of south-east Asia, including the Malay Peninsula, Java and Borneo. It extends west across the Himalayas to Pakistan, east along the south-east coast of China and on Taiwan.

Evolution and relationships See Elegant Scops Owl (page 107) for placement of *Scops*. The Mountain Scops Owl is closely related to other east Asian *Otus* species, many of which are endemic island forms.

Description This is a rather compact scops owl. The upperpart plumage is rufous with dark barring and a white scapular line and covert bar. The underparts are lighter reddish with fine black-and-white bars, as are the nape and crown. The facial disk and pale eyebrows show little definition or contrast; the ear-tufts are small. It has orange-yellow eyes, a horn-coloured bill and grey claws, feathered tarsi and bare toes. White wing barring is prominent in flight. Juveniles are duller.

Geographical variation About nine subspecies are recognised, though there is also marked individual variation. *O. s. luciae* on Borneo is the darkest form, while *O. s. vulpes* of the Malay Peninsula is bright orange-red.

Movements and migration It is mostly resident, with some altitudinal movement.

Voice The male's song is a simple, sweet double whistle, easily imitated, given every second or so for long spells. Females respond with a single-note song.

Habitat It lives in damp, dense upland forest, including tropical rainforest in the south of its range.

Behaviour, hunting and diet It is nocturnal and active from dusk. Like other scops owls it eats many large insects. The hunting behaviour is little observed, but it probably hunts from a perch and also chases insects in flight.

Breeding Little is known about its breeding behaviour. Males stay on territory and sing all year round, though their activity intensifies in spring. It usually nests in tree-holes, especially old or abandoned woodpecker nests. The female lays two to five eggs, which she incubates alone while the male brings food to the nest for her and, in due course, the nestlings.

Status and conservation Though reported as common in some areas, it is threatened by logging activities. Its population is not known but as its range is large and its numbers apparently stable, it is evaluated as Least Concern.

Its inhospitable forest habitat means that the Mountain Scops Owl is little studied.

Oriental Scops Owl
Otus sunia

A typical small scops owl, the Oriental Scops Owl is active by night and seeks concealed roosts in the daytime.

Size 16–19cm

Range This is an east Asian species, found in Vietnam, Cambodia, Thailand, east China and north to Korea, Japan and south-east Russia including Sakhalin. There are separate populations in west and south India and Sri Lanka.

Evolution and relationships See Elegant Scops Owl (page 107) for discussion of *Otus*. This species is closely related to other Old World scops owls, including Scops and Pallid Scops Owls.

Description This is a small, slim, large-headed owl that occurs in rufous and grey morphs. The plumage is a typical scops owl combination of complex streaks and barring on the upperparts and cross-barred black streaks on the underside, in shades of grey and brown (red-toned in the rufous morph) that from a distance look rather plain and uniform. The facial disk is not clearly defined. It has pale eyebrows and quite large ear-tufts that may be erect or flattened. The eyes are yellowish-orange, the bill and claws grey, the tarsi lightly feathered and the toes bare.

Geographical variation About seven subspecies are recognised, with southern forms in general darker, redder and smaller than the more northerly ones. The very richly coloured *O. s. malayanus* of the Malay Peninsula may warrant full species status.

Movements and migration Northerly birds move to India or the southern Malay Peninsula to winter.

Voice The song comprises one long and two shorter notes, slightly rasping but with the ringing quality of water drips, repeated for long spells. There is some variation in song form and tone between subspecies.

Habitat It inhabits light or open woodland, including town outskirts in some areas.

Behaviour, hunting and diet It is nocturnal, roosting in trees or holes through the day. It hunts from a perch and in flight, taking mainly insects and other invertebrates, but also small mammals, reptiles and birds.

Breeding This species' breeding behaviour is not yet well studied, but is probably very similar to that of other scops owls.

Status and conservation The population size is not known. It is quite a versatile species, and there is no evidence that the population is declining. This, along with the fact that its range is very large, means it is currently assessed as Least Concern.

Right: This Oriental Scops Owl was photographed in Hong Kong, where it occurs as a rare passage migrant.

Collared Scops Owl
Otus lettia

Size 23–25cm

Range Another south-east Asian species, this scops owl occurs along the Himalayas and then broadly across south-eastern China and Taiwan, and down to Thailand, Laos, Cambodia and Vietnam.

Evolution and relationships See Elegant Scops Owl (page 107) for the placement of *Otus*. This species is closely related to the Indian Scops Owl *O. bakkamoena*.

Description This is a larger scops owl than the previous species, with a complex pattern in browns and greys. The upperparts have barring on the flight and tail feathers and larger coverts, a vague pale scapular line, and streaks on the smaller coverts, mantle, nape and crown. The underparts are brown with black cross-barred streaks. The brown facial disk is outlined with blackish spots, there is a whitish collar and the broad pale eyebrows lead to blunt ear-tufts. The large eyes are dark brown (usually looking black, especially when half-closed); the bill and claws are a greyish horn colour. The tarsi are feathered but the toes are bare. Juveniles are more barred.

Geographical variation There are four or five sub-species. *O. l. plumipes* has feathered toes and darker eyes. *O. l. erythrocampe* of southern China, which may be a separate species, is more chestnut-toned with brighter eyes.

Movements and migration More northerly populations migrate south as far as India and the Malay Peninsula in winter.

Voice Singing males give a single down-slurred hoot, lower pitched than that of the other scops owls described, every 15 to 20 seconds. Females respond with a similar but higher song. In alarm it gives a rapid chatter.

Habitat It uses various wooded habitats from forest and scrub to gardens on town outskirts.

Behaviour, hunting and diet By day this owl sits quietly on a branch or against a tree trunk, becoming active at dusk. It hunts primarily from a perch, swooping down to catch prey on the ground. As well as invertebrates, it will prey on rodents, small reptiles and small birds.

Breeding The breeding behaviour is not yet well studied, though is probably similar to that of other scops owls. Song may be heard at most times of the year. The nest-site is a tree-hole, up to 5m high, and the usual clutch size is three or four eggs.

Status and conservation This species is thought to have a stable population, though deforestation is a potential threat. Numbers are not known but because of its extensive distribution it is classed as Least Concern.

Unusually for a scops owl, this species has dark eyes.

A beautifully patterned small owl, the Elegant Scops Owl is little known and rare.

Elegant Scops Owl
Otus elegans

Size 20cm

Range This rare owl is found only on a few islands in south-east Asia, including Batan and Calayan in the Philippines, Ryuku and Daito off Japan and Lanyu island off Taiwan.

Evolution and relationships The diverse genus *Otus* belongs to the large subfamily Striginae, forming a distinct basal lineage within that clade. The taxonomic position of the Elegant Scops Owl has not yet been studied.

Description This medium-sized scops owl occurs in red and grey morphs. It has a typical complex and subtly toned upperpart plumage of grey, whitish and brown, with barring on the long feathers and dark arrowhead markings on the shorter ones. The under-parts, nape and crown bear cross-barred streaks. The facial disk is well defined, edged with blackish. It has pale eyebrows and short pointed ear-tufts, which may be held flat, yellow eyes, partly feathered tarsi and bare toes. It looks long-winged in flight. Juveniles are similar to adults.

Geographical variation The nominate occurs on the Japanese islands. In the Philippines is the darker *O. e. calayensis*, on Lanyu the slightly paler *O. e. botelensis*.

Movements and migration Northern birds may be partly migratory, southern populations are sedentary.

Voice The singing male gives a three-note croak, the last note longer, repeated every second or two.

Habitat It lives mainly in forests but sometimes in human-modified partly wooded environments.

Behaviour, hunting and diet Its habits are very little known. It is nocturnal, roosting by day in a hidden spot. Like other scops owls it hunts mainly insects and inver-tebrates, which it catches on the ground or in flight, making sorties from a perch.

Breeding The breeding season begins in April, when the male sings to attract a mate (not necessarily the same one as the previous year). The nest-site is a woodpecker hole or other tree-hole, in which the female lays three to five eggs, incubation probably beginning when the first or second egg is laid. Fledging is likely to be at about four weeks old, and independence another four weeks later.

Status and conservation Deforestation threatens this species' habitat. Its population is unknown, but because of its very small range and therefore presumed small population size, this owl is assessed as Near Threatened.

Common Scops Owl
Otus scops

Size 16–20cm

Range This owl is a summer visitor to southern Europe (including most Mediterranean islands) and just into north-west Africa. Its distribution extends narrowly into parts of central and southern Asia, as far east as north-west Mongolia and as far south as north-west India, but is absent from the more arid parts of the Middle East and south-central Asia. It winters in west and central Africa, south of the Sahara but north of the rainforest regions.

Evolution and relationships See Elegant Scops Owl (page 107) for discussion of the genus *Scops*. Of the *Otus* species that have been studied with DNA techniques, the Pallid Scops Owl is the Common Scops Owl's closest relative, but there are other Old World *Otus* species, in particular the widespread African Scops Owl *Otus senegalensis*, that are likely to be equally or more closely related.

Description A small, large-headed and often slim-looking owl, the Common Scops Owl has beautifully camouflaged plumage patterned like tree bark, strikingly similar to that of certain unrelated bird species such as the Wryneck *Jynx torquilla* and European Nightjar *Caprimulgus europaeus*. The upperparts have a complex pattern of bars and streaks in shades of grey, sandy and brown, with a whitish scapular line. A redder morph with richer brown plumage occurs occasionally. The underparts are light grey-brown with fine, cross-barred black streaks. The nape and crown bear a fine pattern of streaks and barring.

The facial disk is clearly but finely outlined with blackish, and is marked with very fine concentric dark circles on a pale background. The broad pale eyebrows extend up to the large (but sometimes held flattened), blunt-tipped ear-tufts. The eyes are yellow, with prominent upper eyelids that give it a sombre expression. The bill and claws are greyish, the tarsi are feathered but the toes are bare. In flight it looks long-winged. Juveniles have fluffier-looking plumage with more prominent fine barring.

Geographical variation Five or six subspecies are recognised. The nominate is found in the western part of the range, as far as north Turkey. The form *O. s. turanicus*, which is paler and greyer, extends from Turkmenistan to the west of Pakistan, and another greyer form, *O. s. pulchellus*, is found in the far east of the range. On the Balearic Islands and southern Spain is the pale *O. s. mallorcae*. The darkest form is *O. s. cyprius*, of Cyprus and Turkey.

Movements and migration Unusually among owls, this is in general a true migrant, with predictable and widely separated summer and winter ranges. They arrive on their breeding grounds as early as March (later

The appearance of this bird's face varies greatly depending on its state of alertness.

Left: The intricate grey pattern helps the small Common Scops Owl camouflage itself among the foliage by day.

Fledglings that have just left the nest will wait in nearby branches for their parents to bring food.

further north) and depart between August and November. However, a few more southerly populations are resident. Occasionally a bird returning from Africa in spring may overshoot its breeding grounds, especially if weather conditions are fine and warm with southerly winds, and end up hundreds of kilometres further north than expected. There are several records from northern Europe, mainly involving individuals trapped in mist nets at bird observatories, and males that establish territories and draw attention by singing for a mate.

Voice The male's song is a steady series of single piping or mewing notes, delivered every two or three seconds for very long spells. The notes are slightly down-slurred with a rather plaintive tone, like a cut-short gull call. The female responds with a similar but higher-pitched song, and the pair will duet, timing their notes alternately. Other calls include a loud yelping alarm call, and a soft 'phew' contact call between male and female.

Habitat This owl prefers quite open, warm and dry landscapes with scattered trees and bushes, at both high and low altitudes (up to 3,000m in the east of its range). It will inhabit farmland, small copses, olive groves, scrubby areas, large gardens and the edges of towns. In its winter quarters it is found in savanna with scattered trees, and bushy scrub.

Behaviour, hunting and diet The Scops Owl is nocturnal, active mainly in the first period of the night between sunset and midnight. By day it makes full use of its camouflage by sitting in an inconspicuous roost, often the same spot day after day. This may be in a tree against the trunk, a hole in a wall or in a rocky crevice, adopting a slim upright posture which becomes more exaggerated if the bird is disturbed.

This owl hunts mainly by watching from a perch and then gliding or dropping down onto its prey. It will also hawk insects in flight. The diet is mainly large insects and other invertebrates, especially those active at night – moths, cicadas, beetles and crickets make up a significant proportion of its food. It will wait near street lights to hunt the insects that are attracted to the beam, either catching them in flight or picking them off walls and other surfaces. It also catches small vertebrates – rodents, lizards, frogs and the occasional bird. Small prey will be eaten in one go but it will hold larger prey in one foot and lift it to the bill to take bites out of it, parrot-fashion.

Breeding Males and females that have already bred successfully will seek to occupy the same territories as previous years, and so may end up with the same mate for two or more breeding seasons although their tie is to the territory and they will happily accept a different mate.

The male's incessant territorial call begins as soon as the bird has returned from migration (or as early as January or February in resident populations). Males that fail to attract a mate will continue to call for hours every night well into summer, but once paired up they call much less. When a female joins the male, the pair will duet and copulate, and the male will show the female a nest-site by entering it and singing from it. Occasionally a male will pair with two females and provision them both in separate nest-sites.

It will nest in all kinds of cavities, depending on what is available in the habitat. Tree-holes and woodpecker holes, cracks in walls or rock faces, cavities in the rooftops of old buildings, nestboxes and even well-hidden animal burrows in the ground are all potentially suitable. A good site will be used for successive years, but birds that lose their eggs or chicks will seek a new nest-site the following year.

The female usually begins to lay in late April or May, and produces between two and six eggs (three or four is most usual). They appear at two-day intervals, and as is usual with scops owls she starts to incubate when the second one is laid. The incubation period is usually about 25 days but can vary a few days either way. Incubating alone, the female relies on her mate to bring food to her. He roosts nearby during the day. Once the eggs hatch the male increases his rate of food delivery, while the female remains with the chicks until they are two or three weeks old. After this, she joins the male in hunting through the night to feed the growing brood.

The chicks are quite mobile on their feet by four weeks old and leave the nest, scrambling to safe hiding places in trees or bushes or among rocks nearby, though it will be another week or so before they are able to fly, and another four weeks before they are independent. They will have only a few more weeks at most before beginning their first migration, and the following spring they will return and be ready to breed.

Status and conservation The Scops Owl is common in some parts of its range (especially around the Mediterranean) but scarce in others. Overall, it is thought to have a total population of between 1 and 3 million individuals. Its dependence on larger insects makes it vulnerable in areas where agricultural practices are changing, with increased use of pesticides and removal of patches of natural 'rough ground' reducing the numbers of this kind of prey.

It may also suffer locally from a lack of nest-sites and, in some areas, is threatened by increasing populations of other owl species. For example, more Tawny Owls are breeding in the Camargue in France and are preying on Scops Owls. A gradual decline appears to be under way across much of its range, but because of the size of that range and the generally large population it is currently assessed as Least Concern.

This well-grown chick is already a competent climber and will achieve full independence in a few weeks' time.

Pallid Scops Owl
Otus brucei

Size 16–20cm

Range This owl has a fragmented distribution, with at least six isolated populations in mainly coastal parts of the Middle East including the UAE and northern Oman, Israel, Palestine and parts of Syria, Iraq and Iran, and a separate population extends east through central Asia into north-west India, China, Pakistan and Afghanistan.

Evolution and relationships See Elegant Scops Owl (page 107) for placement of the genus *Otus*. DNA study has shown that the Pallid Scops Owl is closely related to the Scops Owl; it was formerly considered conspecific with it but the fact that the two live side by side in some areas supports their separation, as does the DNA evidence and differences in vocalisations.

Description This is a small and slim scops owl with rather long wings and tail. The upperparts look sandy grey-brown from a distance, but bear an intricate pattern of barring (on the longer coverts, flight feathers and tail feathers) and dark central streaks on the smaller coverts and mantle feathers. The crown and nape are speckled dark. The pale sandy belly is marked with dark cross-barred streaks. The facial disk is outlined with a narrow light then dark surround. It has quite broad pale eyebrows which slant upwards to the pale, blunt ear-tufts. The large eyes are orange-yellow, the bill and claws grey. The tarsi are lightly feathered and the toes bare. In flight it looks long-winged. Juveniles have dark upperparts with pale barring, and pale undersides with fine dark barring.

Geographical variation There are between two and four recognised subspecies. The nominate forms the largest single population and the most easterly, in central Asia. The paler form *O. b. exiguus* occurs in Iran and Iraq and further east.

Movements and migration The most northerly central Asian birds are migratory, wintering mainly in north-west India. Otherwise it is resident and sedentary.

Voice The male's territorial song is a series of pumping medium-pitched single hoots, at closely spaced intervals, likened to the coos of a pigeon. Other calls include a deeper single hoot from both sexes, and in alarm various barking or rattling calls.

Habitat This is a bird of rather arid, rocky and sparsely vegetated landscapes, though it requires at least a scattering of trees and bushes. It may occur on the outskirts of towns, along riverbeds, wadis and in the hills.

Behaviour, hunting and diet It is mainly nocturnal but sometimes active by day, especially in the morning and evening. It chooses concealed spots in hollows among rocks or in bushes as its daytime roosts. When roosting it adopts a slim and erect posture, with the eyes closed to slits, the facial disk narrowed and the ear-tufts fully raised, giving a distinctive expression which can change dramatically to a wide-eyed and round-faced look when alert. It usually uses a sit-and-wait hunting method, sitting on a high perch ready to swoop down on passing prey. It also chases flying prey.

The diet is mainly composed of insects, such as moths and beetles, and other arthropods like spiders and scorpions. Other prey includes lizards, small snakes, birds and small mammals including bats.

Breeding The male may begin singing as early as January, though courtship may be quite prolonged. When a female joins him, the pair establish their bond through duetting and mutual preening, and the male leads her to his chosen nest-site, singing from its entrance. The nest-hole may be an old woodpecker nest or natural tree-hole, but where trees are in short supply it will use holes in walls, rock crevices, gaps between boulders or other birds' nests. The female will lay four to six eggs, beginning to incubate alone after the second while the male keeps her supplied with food.

The eggs take about 21–25 days to hatch, and the chicks are guarded by their mother for a week or so before she begins to leave them to assist the male with hunting. They leave the nest as early as three weeks but are not capable of flight until about five weeks, though they are competent climbers and can hide themselves effectively. They are independent at about nine or ten weeks, and ready to breed the following year.

Status and conservation This owl is common in parts of its range, but its fragmented distribution may be cause for concern in the future. Its numbers are not known but are not thought to meet the criteria for conservation concern, so it is categorised as Least Concern.

Right: The Pallid Scops Owl shows a typical scops owl plumage pattern, adapted to resemble tree bark.

Flammulated Owl
Psiloscops flammeolus

Size 16–17cm

Range This owl has a very patchy distribution down the western side of North America, from Canada into Mexico. Its northern limits are reached in British Columbia, and it extends patchily south through Washington, Oregon, Nevada and Arizona as far as Vera Cruz, but does not generally reach the west coast except in the northernmost part of its range.

Evolution and relationships The Flammulated Owl is the only species in the genus *Psiloscops*, which is part of the subfamily Striginae. The species was formerly classified in the screech-owl genus *Megascops*, which was itself formerly subsumed within *Otus*, the Old World scops owls. DNA study shows that separation of the Flammulated Owl at genus level is warranted. *Psiloscops* is more closely allied to *Megascops* than to *Otus* but is clearly distinct from both. Similarities in the rhythm of its song to that of the Scops Owl appear to be coincidental.

Description This is a charming small owl with a large, round head, long wings and a short tail. Even for its size it has a rather small bill and talons, which reflect its insectivorous diet. It occurs in grey and rufous morphs, and intermediate forms with much minor variation over different parts of its range.

The upperparts are marked with an intricate pattern of streaks (on the mantle and inner coverts) and barring (on the longer feathers) in tones of brown, grey, off-white and blackish, with a pale scapular line. The underside is pale but heavily marked with cross-barred streaks. The nape and crown are similarly marked. Birds from the north of the range, which migrate, have noticeably longer wings than the sedentary birds from further south.

The broad facial disk, quite reddish in the rufous morph, is marked with fine concentric circles of dark and light, and is well defined with a strong dark border. The whitish eyebrows extend up to the very short, pointed ear-tufts at the head corners. The tufts are virtually invisible on a relaxed bird. The large eyes are very dark brown or black – this, together with the shape and colour of the facial disk, gives the bird almost the look of a miniature Tawny Owl. The small bill and claws are grey. The tarsi are feathered, the toes bare. In flight the relatively long and strongly barred wings are noticeable. Juveniles are neatly barred on both upper- and underparts.

Geographical variation Although historically up to six subspecies have been recognised, most authorities today reckon that this is primarily individual variation and consider the species to be monotypic.

Movements and migration In the north of its range this owl is a summer visitor. It is not present in Canada at all in winter and is thought to move out of all, or almost all, of the US as well. Additionally, those that nest at the highest elevations undertake altitudinal movements in winter. It shows a rapid movement northwards in spring, as birds hurry to secure the best breeding territories and/or mates, but autumn migration is more protracted, as many of the migrants are inexperienced juveniles. Autumn vagrants have strayed as far east as Florida.

Voice The male's song is a simple, somewhat clipped hoot or coo, rather low pitched for the bird's size and almost the same tone as a toy trumpet. The Flammulated Owl has an unusually thickened syrinx and swollen throat skin, apparently associated with its unusually deep voice. The single note is repeated every two or three seconds, or may be given as a double call when the owl is excited. In alarm it may produce a high-pitched mewing note or an alarming

Left: Barely larger than its pine cone perch, the Flammulated Owl is one of North America's smallest species.

The stubby ear-tufts are erected in its camouflage posture.

Right: Resting in a tree hollow, this Flammulated Owl is beautifully camouflaged against the bark.

screech. Begging chicks make a wheezy hissing sound and a rapid twitter.

Habitat This owl is a species of open upland woods composed mainly of pines and firs with a scrubby understorey, in warm and dry situations. It is closely associated with areas of Ponderosa Pine *Pinus ponderosa* at lower elevations. It breeds at altitudes of up to about 3,000m in the more southerly parts of its range, but as low as 390m in some areas. It prefers woodlands with a generous understorey.

Behaviour, hunting and diet The Flammulated Owl is nocturnal, though feeding activity appears to be more intense just after dusk and just before dawn. By day it roosts pressed close to a tree trunk, standing slim and upright in a camouflaged posture if disturbed. By night it moves around from tree to tree, searching for prey among the foliage, and some is taken in flight. Flying insects are chased in the air and caught with the bill rather than the feet. Its habit of foliage-gleaning is highly unusual among owls, and this together with its way of catching insects in flight is more reminiscent of an insect-eating songbird than a bird of prey. It is quite agile in flight and very active when searching for food.

It is almost entirely insectivorous, taking a variety of larger insects and other arthropods such as moths, crickets, grasshoppers, beetles, spiders and scorpions. One study found the diet comprised 72 per cent grasshoppers, crickets and related species of the insect order Orthoptera by number, with the remainder a much more diverse collection of other insect and arthropod groups. Only a very few records of it taking vertebrate prey are known – vole remains have been found in pellets and small songbird feathers at a nest.

Breeding While pairs may stay together through the year in areas where it is sedentary, the northern migrant population must re-establish territories and find mates anew each season, and often will have a different mate and territory to the previous year. Males sing from mid-spring to attract a female and draw her attention to potential nest-sites. These are usually nest-holes excavated by flickers or other woodpecker species, or natural cavities; they will also use nestboxes. Holes as small as 4cm across are accessible to this species, and using smaller holes helps to avoid attack by certain nest predators.

The female usually lays three or four eggs at two-day intervals, and begins incubating after the first or second is laid. The male makes very frequent food deliveries to her – mainly of moths – while she sits for the three-week incubation period. She broods the young owlets for a further week or so before adding her efforts to the male's in bringing food for the family. With often very small prey being brought singly to the nest, the feeding period places a heavy workload on the parents. The chicks leave the nest at about three weeks old, while still clad in down and unable to fly strongly, and disperse into nearby trees. They will remain within 100m of the nest-site for at least a week.

Within about three days of fledging, the brood is divided, and the male begins to feed one or two chicks exclusively while the female cares for the other one or two. This continues until the youngsters are independent, about five weeks after fledging. The chicks suffer heavy predation post-fledging, being small enough to make an easy meal for practically every predatory bird and mammal that lives in the same habitat. Mastering concealment behaviour is of as much importance as learning to find food for these vulnerable chicks. Should they survive the challenges of the winter, including migration, the following year they will be old enough to breed.

Status and conservation An elusive species, this owl presents plenty of challenges for anyone wishing to study it. It has an extensive range, and an estimated total world population of 40,000 birds. Numbers appear to be gradually declining in the north of its range at least. It is vulnerable to habitat loss from logging, as its preferred habitat unfortunately is forest that is often commercially valuable. It is also at risk from the effects of pesticide use, depending as it does on a good population of larger and flying insects.

It has a low reproductive rate, especially for a small species, which means it can be slow to recover from declines. In some US states it is a species of conservation concern, with conservation measures under way, including efforts to fully survey its population and the way it uses its habitat. Providing nestboxes is an effective way to increase breeding populations locally. On a global scale, the species is currently assessed as Least Concern.

Western Screech Owl
Megascops kennicottii

Size 22–24cm

Range This owl is distributed in a broad band down the western side of North America, from the southern tip of Alaska down to central Mexico, and east as far as western and north Texas.

Evolution and relationships See Eastern Screech Owl (page 123) for discussion of the genus *Megascops*, and the genetic separation of Western and Eastern Screech Owls. These two birds not only show sufficient DNA convergence to warrant separation into full species, but also have distinctly different vocalisations, and in the few areas where their ranges overlap (for example, at the Pecos River in Texas) they do not normally interbreed (although there are a handful of records of hybridisation in the wild). However, their morphological and ecological similarities do show what close cousins they are.

Description The Western Screech Owl is small and large-headed, and like the Eastern Screech Owl is more robust-looking than the *Otus* scops owls of the Old World. It occurs in a grey morph and a scarce red-brown morph, and in general shows considerable variation in plumage tone, with populations inhabiting more humid areas tending to be darker overall. Northern birds appear to be slightly longer-winged than their southern counterparts. Apart from by range, it is difficult to separate from the Eastern Screech Owl, given that both species are quite variable in plumage, but at close range it can be seen to have a darker bill, with a small pale tip.

The plumage overall is very similar to that of the Eastern Screech Owl, with streaked mantle and inner coverts, a pale scapular line, and barring on the outer coverts, tail and flight feathers (more pronounced and contrasting on the primaries). The underparts have heavy streaks with pronounced cross-barring, thinning out towards the centre of the breast and belly. It has a white chin patch. The nape and crown have fine dense barring with cross-streaks.

The facial disk is heavily outlined with black on the outer edges but open at the bottom. The pale eyebrows are rather narrow, leading up to the neat ear-tufts. The eyes are yellow or orange-yellow, the quite powerful-looking bill is blackish and the claws grey. The tarsi are feathered but the toes barely so. The wings look strongly banded in flight, and the bird looks top-heavy and short-tailed. Juveniles have more barring than adults.

Geographical variation Some eight subspecies are usually recognised, though more study is needed to properly determine their validity. The nominate ranges from Alaska south along the coast, perhaps just to the

The Western Screech Owl is very similar to the Eastern Screech Owl, though the two have little overlap in their distribution.

Left: The Western Screech Owl takes a varied diet, including plenty of invertebrates.

Eastern Screech Owl
Megascops asio

Size 18–23cm

Range This owl occurs in the US and the far south and south-east of Canada. Its range extends west as far as Montana and Wyoming, and south into Mexico, and it is replaced by the Western Screech Owl and Whiskered Screech Owl further west and south, respectively.

Evolution and relationships The genus *Megascops* is part of the subfamily Striginae. The screech owls have only recently (the late 20th century) been separated into their own genus, distinct from the Old World scops owls (genus *Otus*). DNA evidence revealed that the New World species form a separate lineage, not especially closely related to *Otus* – their most recent common ancestor lived between 6 and 8 million years ago. The Eastern Screech Owl is closely related to the Western Screech Owl, and the two are still regarded as subspecies of the same species by some authorities. However, DNA evidence supports their separation.

Description A petite, large-headed and short-tailed owl, the Eastern Screech Owl looks similar to the Old World scops owls, though is rounder-faced. It occurs in grey and red morphs. The upperparts are grey (or reddish) with fine dark streaks on the mantle feathers and inner coverts, and a strong whitish scapular line. The outer coverts and secondaries are subtly barred, while the primaries have much bolder dark and light barring. The crown and nape have fine dark streaks. The underparts have a grey-brown or red-brown ground colour, whitening towards the centre of the breast and belly, and marked with quite strong cross-barred dark streaks.

Far left: Guarding its nest-hole, a grey morph Eastern Screech Owl is beautifully camouflaged.

Left, above: This species is mainly nocturnal, so more often heard than seen.

Left, below: Strong legs and feet enable it to easily scramble up a tree trunk, with some wing assistance.

Owlets leave the nest before they can fly, relying on their climbing skills to stay safe.

The sides of the facial disk are strongly outlined blackish but 'open' at the bottom. The disk itself is finely barred in concentric circles. A pale central cross frames the inner side of the eyes and extends upwards into a pair of off-white eyebrows. The top head corners bear blunt ear-tufts, which vary in apparent size depending on the bird's state of excitement. The eyes are yellow, the bill and claws a greyish horn colour. The tarsi and toes are feathered. In flight the banded primaries are striking. Juveniles have subtle barring rather than strong streaks on the underparts.

Geographical variation Somewhere between six and nine subspecies have been recognised, though the species' different colour morphs confuse the picture and some forms may just represent individual variation. The nominate is found in the south-east of the range. The form *M. a. floridanus*, restricted mainly to Florida, is more richly coloured, with red morphs particularly common. In the north-easternmost part of the range is the largest, palest form, *M. a. naevius*. The southern-most subspecies is *M. a. maccalli*, a dark and heavily marked form.

Movements and migration This is a resident and sedentary species. Birds in the far north of the range may be forced to move away from their territories to escape severe winter weather, but there is little evidence of significant movement. Dispersing juveniles usually settle within 15km of where they were born.

Voice The male's territorial song is a fast, high-pitched and sweet-toned trill, of some 20 notes given in rapid succession, the pitch and loudness falling towards the end of the trill. In excitement and when duetting with the female, the song becomes shriller, more reminiscent of a horse's whinny. The female produces a similar song in response. The various alarm calls include a harsh yelping or whining note, and chicks beg with harsh hisses. This species and the other *Megascops* owls are said to have a particularly wide range of calls. However, its extensive repertoire does not include sounds that could really be described as 'screeches'.

Habitat It requires trees for nesting but is not a bird of dense, closed forest. It will use all kinds of open woodland but is especially drawn to low-lying and rather damp areas with a good mix of habitat types, including scrub, fields, riversides and hedgerows. It quite commonly nests in suburban areas where there are large gardens containing plenty of mature trees. It prefers deciduous to coniferous trees. There is evidence that the grey morph survives better in the north of the range, presumably because its plumage is better camouflaged among more northerly tree species. The red morph (which appears to be genetically dominant over grey) does better and is more numerous further south.

Behaviour, hunting and diet This is a nocturnal species, spending the daylight hours in a concealed roosting spot, which may be on a branch against the tree trunk or in a tree-hole or thick foliage. It becomes active only after sunset, and so is very easily overlooked even where it lives among human habitation, its night-time song often the only sign of its presence.

When hunting, it will select a high branch with good visibility to the ground and search for prey with both eyes and ears. It moves on to a different perch periodically, and when it finds prey it swoops down, usually seizing the victim on the ground. Small items will be carried off elsewhere but large prey is at least partly eaten on the ground. Some prey is cached for winter, and it also ups its feeding rate in autumn, laying down fat stores for the cold months ahead.

Unlike the scops owls, which are primarily insectivorous, screech owls take a highly varied range of prey, exploiting whatever is most abundant at the time. It will readily take invertebrates including snails, crayfish, spiders, worms and assorted insects. All small mammals are potential prey, including the expected mice and voles but also moles, bats and flying squirrels. Bird prey may be as large as American Woodcock, Feral Pigeon and Northern Bobwhite quail. It has also shown itself to be an intraguild predator, capable of taking American Kestrels. It will also catch lizards and snakes.

Breeding This species forms lasting pair bonds, and the nest-sites of pairs appear to cluster into loose colo-

nies, forming miniature 'neighbourhoods'. Although it may pair for life, individual birds rarely live more than five years due to the many natural dangers that they face, but when a bird loses its mate there is almost always another unpaired bird living nearby and ready to step in. Some studies suggest that females prefer smaller males to larger ones, perhaps because this would mean that between them the pair could efficiently hunt a wider range of prey.

Pair bonds are established (or reaffirmed) from mid-winter. The pair duet together and also spend much time tenderly preening each other's head and neck plumage. Activity increases into spring, when the female accepts a nest-site (more than one potential site may be shown to her by the male). This is usually a tree-hole of some sort – nestboxes are readily used. In it she lays a clutch of between one and eight eggs, the average clutch size ranging from three in the south-east of the range to five in the north. Incubation begins with the first egg, with the next coming two days later and so on until the clutch is complete. She incubates alone while the male brings food. Interestingly, a particular prey type – blind snakes of the genus *Leptotyphlops* – is often delivered to the nest live. These snakes will live in the nest long term, feeding on the various parasites and scavenging insects that naturally proliferate as prey remains and droppings build up inside the cavity.

After 26 days of incubation the clutch begins to hatch. The female stays with the nestlings, brooding them and carefully feeding them scraps of food. As they grow older, they cease to require brooding and can swallow smaller prey whole – by the age of two weeks they are left alone in the nest while both parents hunt for them. They leave the nest at four weeks of age, already able to fly. Over the next two or three weeks their flight becomes stronger and they learn to hunt for themselves, taking mainly small and slow prey at first. Most will make their first breeding attempt the following year.

Status and conservation This owl is adaptable and tolerates the presence of humans in its habitat, factors that may explain its current period of increase, as more birds move into suburbia. Those in more rural surroundings are at risk from land-use changes and more intensive farming practices. However, a general trend of gradual increase has been evident since the 1970s. This, together with the species' large distribution and a sizeable population of some 800,000 birds, means that it is assessed as Least Concern.

The red or rufous morph is most abundant in the wetter and warmer parts of the species' geographic range.

When alarmed, the Eastern Screech Owl will adopt a tall and sleek concealment posture.

Whiskered Screech Owl
Megascops trichopsis

Size 17–19cm

Range This owl is found through much of the Mexican uplands, especially the west. Its range extends south through highlands in central and western parts of Central America into Panama, and north into south-east Arizona.

Evolution and relationships See Eastern Screech Owl (page 123) for discussion of the placement of *Megascops*. This species is not especially closely related to the two more northerly *Megascops* species already described, but forms a lineage with two other South American species, the White-throated Screech Owl *M. albogularis* and Tropical Screech Owl *M. choliba*.

Description It is smaller than the Eastern and Western Screech Owls, and looks compact and almost squat with a proportionately large head. Grey and red morphs occur. It gets its name from the particularly long and hair-like feathers around the base and sides of the bill. The pattern is similar to that of the other screech owls, with streaks on the shorter upperpart feathers and barring on the longer ones, a white scapular line and particularly strongly banded primaries. The underpart streaks are longer and thicker further down the belly, and the head and crown bear short, cross-barred streaks. The sides of the facial disk are marked with a fine dark line, opening out at the bottom. The ear-tufts are short and may be invisible. It has yellow eyes and feathers on the tarsi which thin out halfway along the uppersides of the toes. The bill is a greyish horn colour and the claws grey. Juveniles are barred all over.

Geographical variation There are three subspecies. The nominate is found in central Mexico. To the north is *M. t. aspersus*, a slightly smaller form which does not occur in a red morph. The most southerly subspecies is *M. t. mesamericanus*, which is more delicately marked, and its red morph is duller than the nominate red morph.

Movements and migration It is resident and sedentary. There may be some altitudinal movement at higher elevations.

Voice The territorial song is a quite fast series of about nine equally spaced single hooting notes, descending the scale towards the end. It takes about two seconds to complete, with a pause of several seconds between phrases. In courtship and when duetting it gives a more rhythmically varied song, and the female's reply is higher pitched. It has a mewing contact call.

Habitat It inhabits dense mountain forests of oak and pine, and also coffee plantations, occurring up to 2,500m and rarely lower than 1,500m.

Behaviour, hunting and diet It is nocturnal, roosting in thick cover until dusk. It finds most of its prey by working its way through the branches and picking off insects or hawking them in flight. Less often, it will still-hunt from a perch.

Unlike the larger screech owls, this species' diet is composed almost entirely of invertebrates. Flying insects taken include moths, crickets, beetles and mantids, while it also picks spiders, millipedes, caterpillars and centipedes from leaves and branches. Vertebrate prey has rarely been recorded, but it has been known to take the occasional small rodent.

Breeding The larger screech owls show considerable tolerance to their conspecific neighbours, but the Whiskered Screech Owl is highly territorial, defending an area of about 300 square metres. Pairs remain together all year and long term, with courtship involving prolonged mutual preening sessions and prey gifts from male to female.

Little is known about the specifics of its breeding behaviour. The female begins to lay usually some time in the second half of April, producing a clutch of three or four eggs in a tree-hole nest. She begins incubating with the first, and is fed on the nest by her mate throughout incubation. The incubation period is not known but is likely to be in the region of 22–26 days, comparable with other screech owls. The same goes for the chicks' maturation and fledging period.

Status and conservation The forests this owl uses are not under high pressure from human activity at present, and its population is thought to be generally stable or increasing in its Mexican and US range. Its total population is placed tentatively at 200,000 birds. With this population across a fairly extensive range and no evidence of decline, it is assessed as Least Concern.

Whiskered Screech Owl is smaller than the North American screech owls and takes mainly insect prey.

Left: Less well studied than its northern relatives, the Whiskered Screech Owl is an enigmatic species.

Snowy Owl
Bubo scandiacus

The Snowy Owl is beautifully adapted in both anatomy and behaviour to its life in the high Arctic.

Right: The feet and toes are fully and thickly feathered for essential insulation.

Size 52–66cm

Range This is an owl of the high Arctic, breeding all the way across Eurasia and North America at latitudes north of 55°N, and wintering in a broad band south of the breeding area to the central US states and in Asia into Kazakhstan, Mongolia and the far north of China.

Evolution and relationships By appearances alone, this species seems to be a highly atypical member of the genus *Bubo*, sometimes placed in a separate genus *Nyctea*. However, DNA studies place it squarely within *Bubo*, with the Great Horned Owl its closest relative. The analysis shows that these two species shared a common ancestor as recently as 4 million years ago, making them more closely related than many superficially more similar species pairs. The Snowy Owl's very obvious morphological differences from the other *Bubo* species appear to represent relatively rapid divergent evolution due to its unique habitat.

Description This is a large and stocky owl with a proportionately small head, similar in shape to other *Bubo* owls but lacking visible ear-tufts. The species is unusual not just in being predominantly pure white, but also in showing clear sexual dimorphism with adult males and females distinctly different in plumage. The adult male is almost immaculate white, with just a few flecks of black here and there on the crown or nape and on the coverts and flight feathers. The female is also white but much more heavily marked, with black bars across the covert,

flight and tail feathers, scattered black flecks on the crown and nape, and a black-barred breast and belly. In both sexes the facial disk is pure white with no defining margin. The face is heavily feathered, obscuring all but the bill tip, and in profile looks more pointed than that of most owls. The eyes are yellow and often look narrowed; the bill and claws are blackish. The tarsi and toes are thickly feathered. In flight it looks quite long-winged. Recently fledged young birds have dark grey downy feathering around the head and body, while older juveniles are heavily barred with black, especially young females, to the point where they look grey from a distance.

Geographical variation This is a monotypic species.

Movements and migration The majority of Snowy Owls migrate south to avoid the Arctic winter, some travelling considerable distances. The migration is rather unpredictable, with irruptions occurring in some years that result in individuals turning up beyond the usual winter range, and on isolated islands. These irruptions are associated with peak lemming years when there are more young Snowy Owls competing for winter territories. Young males travel further than young females, and the same is true of adults, with the larger females perhaps better able to defend the first productive feeding grounds they encounter and thus avoid a longer migration.

Voice As with other *Bubo* species, the male Snowy Owl has a hooting song, though the hoots have a harsh

quality. Up to six are delivered in an uneven sequence. The female may join in with a similar but higher-pitched song. The species gives various contact and alarm calls around the nest, including a thin rasping scream, a coughing or barking note, and a gull-like mewing.

Habitat Nesting habitat is low-lying open and rocky Arctic tundra, a very bare environment where the only vegetation is moss and lichen on the rocks, and there is snow on the ground in places throughout the summer. In some parts of the range it will breed at higher elevations on mountainsides, while in the extreme north it will nest right on the coast. On migration and over winter, the species sticks to open habitats, including moor and grassland, upland river shores, bare islands and even occasionally airfields, playing fields and similar modified open habitats.

Behaviour, hunting and diet The Snowy Owl is primarily diurnal. When not active, it rests on the ground or on top of a rock, making no particular effort to conceal itself. It watches for prey from its perch, and attacks with a long, fast and shallow glide. It will also walk about to search and listen for prey on the ground.

On its breeding grounds this owl depends very heavily on lemmings and voles, which are represented by several species across the Arctic. This is due to a paucity of other prey rather than any inability to tackle larger quarry – on migration and in winter it hunts a diverse range of prey including birds up to the size of ducks and grouse, and mammals as large as hares. In poor lemming years breeding Snowy Owls will hunt ground-nesting birds and their chicks but still suffer greatly reduced breeding productivity.

Like other large owls it is an intraguild predator, taking other birds of prey when it gets the opportunity. On its breeding grounds there is little competition to worry about, but on migration and in winter it will maintain a feeding territory and will hunt and kill other owls – one notable observation involved a single migrating Snowy Owl killing at least three Short-eared Owls (also migrants) on a small island over just two or three weeks. Excess prey may be killed and stored for later use, and the owl will also feed on carrion when prey is short.

Breeding As this migratory species does not always return to the same breeding grounds from year to year, pair bonds are not necessarily maintained beyond a single season, and both polyandry and polygyny have been observed, as well as apparent promiscuous mating with neighbouring pairs. However, a pair that nested in Fetlar on the Shetlands, and did not move away from the island in the winter months, stayed together for nine years.

The breeding grounds are usually occupied in mid-spring, while there is still considerable snow cover. The male advertises his territory with a hooting song but also with conspicuous display flights, involving undulating flight paths and exaggeratedly deep wing-flaps. The performances continue after pairing, as part of courtship. He will sometimes carry prey while displaying, delivering it to the female after the flight, and he will also prepare potential nesting scrapes for her inspection.

The nest-site is usually a slight hollow on the ground in an otherwise exposed situation, affording a good view of the terrain around and allowing the birds to detect approaching danger from a great distance. The female lays a clutch of between three and 14 eggs (more in good lemming years) and starts to incubate as soon as the first egg is laid. Intervals between laying vary from between two and five days, though the hatching intervals are less than this. Nevertheless, the result is often a very large brood spanning a wide age range, the younger chicks providing an insurance policy for the older ones should food shortages occur.

Throughout incubation the male hunts and the female stays on the nest, but both will assist in nest defence – the high visibility factor of their habitat means that a would-be predator such as an Arctic Fox can be challenged well before it is near the nest. The female will use distraction displays to lure predators from the nest,

while males aggressively swoop at intruders.

The chicks, clad in dark down that helps conceal them against a less snowy summer tundra, make their first explorations around the nest-site when they are two weeks old. The female remains nearby to brood and guard them for a week longer before she starts to join in with hunting for the family. The first flights take place at about a month old, and by two months they are flying strongly and approaching independence and the first long flights they will take as they disperse southwards in search of feeding grounds. They will reach breeding maturity at two years old.

Status and conservation The Snowy Owl population is subject to marked fluctuations, in response to the cyclical boom-and-bust population changes of their rodent prey. The owls' reproductive potential is considerable, which means they can build their numbers very quickly when times are good, but accurately assessing the population size is difficult, and there is the risk that any trend independent of prey numbers could be overlooked. The species' most significant threat is that posed by climate change, limiting suitable breeding habitat, and numbers are thought to have declined in North America at least. The population is estimated to be about 300,000 individuals, and its conservation status is Least Concern.

The mesoptile down is not white but grey, to blend in with a snow-free summer tundra.

Great Horned Owl
Bubo virginianus

The Great Horned Owl is the largest owl species across much of North America.

Size 45–60cm

Range This New World owl has a very wide range, encompassing almost all of North America except the far north of Alaska and Canada, and extending continuously down through Central America into the north-west of South America, spreading narrowly south into Peru and east along the coast of Venezuela. A separate population occupies southern and eastern Brazil and south into Uruguay and Argentina.

Evolution and relationships A typical *Bubo* owl (see Pharaoh Eagle Owl (page 140) for discussion of *Bubo*), this is one of only three New World species in the genus. Unsurprisingly, the Great Horned Owl is most closely related to these other New World species, the Snowy Owl *B. scandiacus* and the Magellan or Lesser Horned Owl *B. magellanicus* of western South America.

The latter species is considered by some to be a subspecies of the Great Horned Owl, but DNA evidence supports its separation.

Description This owl is rather variable – the description relates to the nominate subspecies. It is similar to the Old World eagle owls in shape, being thickset, stocky and rather small-headed. The upperpart plumage is patterned in dark and light shades of grey and brown, the mantle feathers bearing pale fringes and dark arrowhead patterns down the centres, and the longer covert, flight and tail feathers barred alternately dark and light. The underparts have a warm mid-brown ground colour and are marked with heavy dark streaks on the upper breast that narrow and spread into dark barring further down onto the belly.

The facial disk is red-brown with a bold black-and-white surround, strong white eyebrows and narrow white outlines to the eyes, giving a stern expression. The crown and nape are finely barred dark, and the large blunt ear-tufts, situated at the corners of the head, are dark-centred. The eyes are yellowish orange, the bill and claws blackish. The tarsi and toes are fully feathered, with fine barring. In flight it looks broad-winged and heavy, the barring on the flight feathers conspicuous on both the upper and undersides. Juveniles are paler and more lightly marked.

Geographical variation There are at least 12 recognised subspecies – genetic analysis is required to fully determine their validity. In general, birds of more forested regions are darker than those inhabiting more open areas. The nominate is found on the eastern side of North America down to Florida. The subspecies *B. v. wapacuthu* is the most north-easterly and northerly form and is very pale and grey-toned. On the western side of North America is *B. v. saturatus*, a dark and very densely marked form with a white collar. A further five or six more possible subspecies occupy the rest of North America, but at least some may in fact represent clinal variation rather than truly distinct subspecies. The South American population is divided into three or more subspecies, including *B. v. nigrescens* of Ecuador and Colombia, the darkest form of all.

Movements and migration It is mainly resident and sedentary but will wander in times of food shortages – this is most likely among the most northerly populations. Young birds disperse a variable distance from their parents' breeding grounds, sometimes more than 200km but usually much less.

Voice Both male and female produce a series of deep, mellow and breathy hoots, each phrase comprising a few notes stuttered together followed by longer single notes. The male's song has three notes in the stuttered phase and two longer notes, while the female gives a four-note stutter followed by a single longer hoot. The

Below left: These owls are powerful and fearless, sometimes swooping at humans who come too near their nests.

Below right: Natural tree hollows large enough for this species are hard to find, but it will use other nest-sites as well.

pair will sing both individually and together in a duet. The usual call is a harsh, low, rasping shriek, while begging chicks make a shriller scream.

Habitat This adaptable species lives in varied habitats across its wide latitudinal range. Wherever it lives it does require the presence of mature trees for nesting and roosting sites. It will breed in coniferous and deciduous woodland, along river valleys, on desert edges, in parks and suburban areas but avoids dense rainforest. It hunts mainly over open ground, so the most suitable habitat offers mature woodland adjacent to farmland, prairie, wet meadows or other open countryside, with proximity to fresh water for drinking and bathing.

Behaviour, hunting and diet It is normally nocturnal, not beginning its evening hunt until after sunset. It roosts by day in a hidden spot in a tree, against the trunk. It favours conifers for roosting, especially in winter when leafless deciduous trees offer less concealment. The roosting posture is tall and upright with the ear-tufts raised.

This owl usually hunts from a perch, choosing a spot at the edge of an open site where it has a wide area to survey and listen for prey – though it appears to be more reliant on sight than hearing. When it detects something it swoops down, usually capturing the prey on the ground. It will also hunt in flight. The prey is usually killed by the sheer force of the grip, but larger or tougher prey may be dispatched by a bite to the head. It hunts most in the hours around sunset and sunrise, becoming less active in the middle of the night.

The prey range taken is very diverse. Mammals predominate, making up as much as 90 per cent by weight, and represented by rodents, rabbits, porcupines, mustelids, opossums, woodchucks and even the occasional raccoon, cat and small dog. There are numerous accounts of it fearlessly attacking dogs in front of their owners. It also takes birds up to the size of ducks or grouse, small reptiles and amphibians, and invertebrates including earthworms and insects. It will readily prey on smaller owl species and raptors. It comes into conflict frequently with the Red-tailed Hawk, a common and widespread medium-sized raptor that shares its habitat in many areas, and while the owl will prey on the hawk, there is at least one documented case of a hawk killing an owl.

Breeding Great Horned Owls are monogamous and form pair bonds that last for successive seasons, as is usually the case with larger sedentary species. The bond between a pair is renewed in late winter, as the male begins to proclaim his territory prior to the start of the breeding season, making the most of the long winter nights. The courtship is conducted mainly via vocalisations between the pair, but the male also offers prey to the female and displays when singing by bowing forwards, tilting up his tail and puffing up his white throat feathers.

The nest-site may be a large tree hollow, but often it is the old nest of a hawk or other large tree-nesting species. Among the Great Horned Owls that nest in more open and arid areas, nests are often built on cliff faces or in hollows on the ground. If a pair nest successfully they will often reoccupy the same site the following season. The female selects the nest and a few days after the first copulation will lay and begin to incubate the first of a clutch of up to four eggs (though just two is the norm). She incubates alone, while the male roosts nearby in the daytime and hunts prey for her (as well as himself) at night. The clutch begins to hatch about 34 days after incubation begins.

The chicks clamber away from the nest as early as four weeks old, waiting in nearby branches and calling noisily for food. They will remain nearby, being fed by both parents, for up to 12 weeks, though they make their first tentative flights from the age of six weeks. They remain reliant on their parents for food for several more weeks, not becoming fully independent until the autumn. Through the course of autumn and winter they will disperse and look for feeding grounds of their own – if they have not moved on by midwinter they will be driven away by their parents as territorial activity for the next season begins. By the time they are two years old they are ready to breed, though may not get the opportunity to secure a territory and mate for another year or more.

Status and conservation This owl is very widespread and adaptable. It is vulnerable to losses from road accidents and, in some areas, deliberate destruction, and may have experienced local declines due to habitat loss and pesticide use limiting its hunting opportunities. However, overall its population is thought to be stable, and to number some 5 million individuals in total. It has a conservation status of Least Concern.

An owlet's wing feathers are well developed some time before it sheds the last of its down.

Left: Different subspecies of Great Horned Owls show different amounts of dark patterning on their undersides.

Eurasian Eagle Owl
Bubo bubo

Size 58–71cm

Range A widespread Old World species, the Eurasian Eagle Owl ranges across south-western Europe and just into the north-west tip of Africa, most of Scandinavia and patchily elsewhere in Europe. Its distribution extends east in a very broad band across central Asia, ranging north into Siberia and south into the Middle East, Pakistan and Burma and east to southern China, Sakhalin and northern Japan.

Evolution and relationships See Pharoah Eagle Owl (page 140) for discussion of placement of the genus *Bubo*. Genetic studies show that the Eurasian Eagle Owl is closely related to other Old World eagle owls, in particular the Cape Eagle Owl *Bubo capensis* and Pharaoh Eagle Owl *Bubo ascalaphus*. The latter species was formerly considered a subspecies of the Eurasian Eagle Owl, and at least one other subspecies may warrant full species status.

Description This is a magnificent and imposing bird by any standards, and is, along with Blakiston's Fish Owl, the world's largest owl species. It looks very bulky, heavy and short-bodied with a relatively small head. The upperpart plumage is coloured with blackish, grey and rufous brown in a complex pattern, with the shorter mantle feathers dark-centred and the coverts, flight and tail feathers barred with dark. The underside has a warm rufous ground colour (though this varies between subspecies). The feathers bear fine dark vermiculations that are evident only at close range, and a strong dark central streak, thicker on the upper breast feathers. This pattern of dark streaks on a warm brown ground colour continues on the nape and crown.

The facial disk is greyer brown, outlined with black, with a broad white central cross centred on the bill base and framing the inner edges of the eyes. There is also a much narrower white outline to the eyes' outer edges. The chin is white, and the ear-tufts are long and pointed, positioned at the top corners of the facial disk and angled outwards. The eyes are a deep orange, the bill and claws blackish. The tarsi and toes are feathered the same colour as the underparts (but without streaks). In flight it looks very large with long broad wings. Juveniles have paler and more diffusely patterned plumage with barring rather than streaks on the underside and smaller ear-tufts.

Geographical variation Many subspecies have been described (more than 12 by some authorities), though some may be invalid and one or more could warrant elevation to full species status. This diversity reflects the species' sedentary nature, meaning that neighbouring populations separated by even quite narrow belts of unsuitable habitat can be sufficiently isolated to develop their own genetic character. The nominate is found in western Europe across to western Russia. In south-west Asia the dark form *B. b. interpositus* is a strong candidate for species status. The Iberian Peninsula is home to *B. b. hispanus*, a smaller, greyer and paler form. In the desert regions of southern central

Calling at night, the owl pumps its tail with each note.

Left: Roosting tight against a tree trunk, this Eurasian Eagle Owl blends in with the background.

The large ear-tufts lie flat when the owl is in flight.

Eurasian Eagle Owls have long, broad wings, and could be mistaken for buzzards at first glance.

Asia are the palest and smallest subspecies, *B. b. omissus* and *B. b. nikolskii*. Further north, subspecies become larger and darker, such as *B. b. yenisseensis* of central Siberia, and *B. b. jakutensis*.

Movements and migration Only the most northerly populations show much movement, making short-range migrations to escape severe weather or prey shortages. It is otherwise a sedentary and resident bird.

Voice The male's territorial song is a very deep, booming single hoot, beginning strongly then downslurring and tailing off. It is given at intervals of some eight seconds. The female also sings, both alone and in duet with her mate, and her song is similar but higher pitched and more disyllabic. In courtship the female solicits food from her mate with a rasping scream, similar to the begging calls of well-grown chicks. Both sexes bill-snap and give a harsh croak when disturbed.

Habitat This is not a forest owl but prefers landscapes with more scattered trees along with rocky outcrops. In different parts of its range it may be found around woodland clearings, quarries, steep hillsides, moorland edges, semi-desert and lowland open landscapes with small copses. In parts of its range it will breed close to human habitation, even within cities.

Behaviour, hunting and diet A mainly nocturnal species, the Eurasian Eagle Owl is active from dawn to dusk. It spends the daytime roosting in a sheltered spot among rocks or in a tree, alone or with its mate. Unusu-

ally among owls, it will sometimes soar and glide on updraughts.

This owl prefers to hunt from a perch, waiting and scanning around for prey, but it will also actively quarter open ground from time to time. It is a fearsome predator, very much at the top of the food chain everywhere it occurs, and is powerful enough to deal with larger prey than any other owl. Its diet is made up of mainly birds and mammals weighing from 200g to 2kg, but it can tackle much larger quarry including foxes, hares and young deer weighing up to 17kg. The largest bird prey taken include gulls and geese. Birds, which may make up more than 80 per cent of the diet by weight, are usually caught by surprise while they are roosting. It is a noted intraguild predator, and there are records of it preying on every other owl species within its range including the Snowy Owl, as well as large raptors such as Goshawks and buzzards.

It kills prey with a squeeze from its powerful talons and bites to the head. Most prey is taken away to be eaten elsewhere, but items heavier than 3kg cannot be lifted and the owl will consume at least part of it on the ground. It will also take carrion when natural food is hard to find.

Breeding The pair bond is long term, often for life, with the two birds defending their territory all year (territory boundaries may overlap with neighbouring pairs). The nest-site is usually on a rocky ledge, but sometimes on the ground below an overhanging bank or bush, or in the old nest of a large bird such as a heron. The same spot will be reused for years if the pair is successful.

The clutch is of one to four eggs, laid at three-day intervals with incubation beginning when the first egg is laid. The incubation period is between 31 and 35 days. However, the eggs' hatching intervals are shorter than the three-day laying interval. Throughout incubation and the chicks' first two weeks of life the male brings food while the female remains at the nest. She deals with the food, tearing it up for the chicks to eat, and defends the site from intruders.

By the age of about four weeks the chicks are strong enough to wander about, though their explorations will be necessarily more limited around cliff-face nests than those on the ground. By this time both parents will be kept busy hunting for them. Fledging occurs at about seven weeks, but parental care continues for another 15–18 weeks. By mid-autumn the chicks will disperse, sometimes encouraged on their way by their no longer doting parents. They reach breeding maturity at two or three years old.

Status and conservation Based on studies of population density in Europe, the total population is estimated at between 250,000 and 2.5 million individuals. The owl's fortunes in Europe are looking fairly rosy with increases recorded in many countries, including Spain and France which hold sizeable populations. However, this upward trend is a recent phenomenon that follows a period of decline, and further east the species is thought to still be declining. The problems it faces include deliberate and accidental destruction by humans, disturbance to its nest-sites and habitat degradation due to development or agricultural intensifica-

tion. However, it remains a widespread and numerous species for now, and overall, it is evaluated as Least Concern.

The case of the Eurasian Eagle Owl in Britain is a special case and a real conundrum for conservationists. Since the late 20th century there has been a small but apparently growing breeding population in England. It is uncertain whether the species has ever inhabited the British Isles historically or whether it has the potential to reach the British coast under its own steam from the nearest mainland populations, but most if not all of the birds now breeding in Britain are almost certain to originate from captivity. There is also a good chance that they include other species of eagle owl, especially the Bengal Eagle Owl *Bubo bengalensis* which is a particularly popular species among bird-keepers. On the other hand, isotope analysis of the feathers of a young Eurasian Eagle Owl killed by a car in Norfolk suggested this individual might well have originated from Scandinavia, which would make it a genuine wild vagrant.

Conservationist concerns about the British owls centre on their possible impact upon Hen Harriers, which share the owls' main breeding grounds in northwest England. Hen Harriers are close to extinction in England, due to persecution, and numbers are too low to withstand additional pressure from the owls, which are notorious predators of other birds of prey. However, while there remains a chance that even one or two of the owls did arrive here naturally or that the species was once native to Britain, removing them will be an unpopular option.

Above left: A part-grown chick, still wearing much of its mesoptile down, can perch and fly short distances.

Above right: Chicks disturbed at the nest assume a puffed-up threat posture to deter attackers.

Pharaoh Eagle Owl
Bubo ascalaphus

Size 45–50cm

Range This eagle owl is found across north-west Africa, including most of Algeria, south into Mali and Niger and extending down the coast as far as Mauretania. A possibly separate population occurs down the Red Sea coast of north-east Africa and across to the Arabian Peninsula.

Evolution and relationships The large genus *Bubo* is part of the major subfamily Striginae, along with other prominent genera like *Strix* (the wood owls) and *Otus* (the scops owls). The Pharaoh Eagle Owl is most closely related to the Eurasian Eagle Owl and has in the past been considered conspecific with it.

Description Smaller than the Eurasian Eagle Owl, this is still a large and solid owl, with a stocky shape and proportionately rather small head. The upperparts are intricately and beautifully patterned with bars and flecks in shades of grey, brown, sandy yellow and black, but from any distance look light sandy grey-brown. The pale yellowish underparts are marked with blackish streaks, becoming longer and finer further down the belly. The pale facial disk is not strongly defined, with just a thin blackish edge. The crown and nape are sandy with dark flecks, as are the wide-set ear-tufts. The eyes are deep orange, the bill and claws greyish. The tarsi and uppersides of the toes are fully feathered. It looks substantial with long and broad wings in flight. Juveniles are similar to adults but plainer, with light barring rather than streaks below and with shorter ear-tufts.

Geographical variation This species is monotypic – paler morphs occur but are found throughout the range and so are not considered to represent a different subspecies.

Movements and migration It is resident and sedentary.

Voice The male's song is a regularly repeated down-slurred hoot, higher in pitch than that of the Eurasian Eagle Owl. Several seconds separate the calls. When pairs duet together, the female's song is higher pitched. Pairs also produce a three-note hoot in courtship duets, with the first note stressed. Other calls are similar to the vocalisations of the Eurasian Eagle Owl – further study is needed to clarify differences.

Habitat This is an owl of arid and sparsely vegetated rocky landscapes, from desert to mountain slope, and savanna with scattered trees on the edges of desert regions.

Behaviour, hunting and diet A nocturnal species, the Pharaoh Eagle Owl roosts by day among rocks and begins to hunt after sunset. Roosting spots may be at ground level or much higher depending on the terrain, and occasionally in trees.

It has a varied diet of mainly the local mammal life, especially gerbils and gundis, but also desert invertebrates, such as scorpions and locusts, snakes and lizards, and birds of various sizes. Larger mammal prey includes Fennec Fox, hares and hedgehogs. It listens for prey from a perch, before attacking with a pounce or gliding flight.

Breeding The pair bond between Pharaoh Eagle Owls is maintained for successive years, and the pair hold territory together all year round. The male's territorial song can be heard all year, but it becomes more intense just prior to the start of the breeding season in late winter.

The nest-site is usually a safe, tucked-away and dark crevice among rocks and consists simply of a scrape on bare rock or ground. Where available it may also use the old nest of another species, or a tree-hole, and even in those most famous man-made structures, the pyramids. A clutch of usually two but up to four eggs is laid and incubated from the first egg's appearance. The female incubates alone, while the male provisions her with food that he brings to the nest. After 31–35 days the eggs begin to hatch, and the chicks are brooded for about two weeks. After this, the female joins in with hunting efforts, and the chicks begin to wander to the nest-crevice entrance, where they may be seen standing out in daylight. They fly at about seven or eight weeks and are independent at about six months but do not reach breeding maturity until they are two years old.

Status and conservation This is a little-known species, although now that it is generally accepted as a distinct species from the Eurasian Eagle Owl interest in it has increased. No clear population trends have been identified. The population size is not known but based on the range is thought to be in excess of 10,000 individuals, meaning that the species is classed as Least Concern.

This beautiful Pharaoh Eagle Owl is an inhabitant of bleak and rocky desert landscapes.

Blakiston's Fish Owl
Bubo blakistoni

Right: This is a very large and imposing owl, about the same size as a Eurasian Eagle Owl.

In winter, Blakiston's Fish Owl has to search for unfrozen stretches of water where it can hunt.

Size 60–71cm

Range This owl has a small, patchy and shrinking world range. It is found in the far east of Russia (including Sakhalin and the more southerly Kuril Islands) and China, and in the north of Hokkaido island in Japan. It may also occur in North Korea, but it has not so far proved possible to determine this because of political tensions making survey work problematic.

Evolution and relationships The three or four fish-owl species, including Blakiston's Fish Owl, are currently classed in the genus *Bubo* by most authorities but were formerly separated into the genus *Keputa*. Genetic analysis shows that the Brown, Buffy (*B. ketupu*) and Tawny Fish Owls form a distinct lineage, but Blakiston's Fish Owl is genetically much closer to the 'core' *Bubo* species. However, all of the fish owls warrant classification in *Bubo*.

Description This is a very large owl and, along with the Eurasian Eagle Owl, is the largest and heaviest in the world. It looks bulky and rather loose-feathered, with a proportionately small head. The upperparts are strongly patterned in shades of grey and brown, with dark centres on the mantle and covert feathers, and alternating dark and light barring with paler tips on the flight and tail feathers. The underparts are paler brown and more uniform-looking, each feather marked with a strong dark central streak (becoming finer further down the belly) and fine dark vermiculations, the latter only evident at close range. The chin is whitish.

The crown and nape are marked with dark streaks, shortest at the front of the crown. The facial disk is poorly defined with no contrasting border, and marked with fine dark streaks radiating outwards, and the prominent bill projects forward rather strongly, producing a less typically 'owlish' face than the typical *Bubo* species. The ear-tufts are located at the corners of the

head, pointing out to the sides, and are rather flat and shaggy. The rather small eyes are orange-yellow, the bill and claws greyish. The tarsi are feathered but the toes are bare, their undersides rough with pointed scales.

Geographical variation The nominate subspecies occurs in Japan, the Kurils and on Sakhalin. The larger, paler and more rufous form *B. b. doerriesi* is found on the mainland – it has a distinctive whitish crown patch.

Movements and migration It is mainly resident and sedentary. Some northern birds appear to leave their breeding grounds in winter but where they go is not known – short-range movements to find suitable feeding conditions seem to be most likely. As long as some stretches of river remain accessible to them they will stay on their territories.

Voice The deep, booming song is commonly performed by a pair as a duet and differs in structure between the two subspecies. In the nominate, the male sings a double note and the female follows with a single hoot, but in *doerriesi* the duet comprises four notes, the first two shorter and softer, the second two longer and louder, with the first and third sung by the male and the second and fourth by the female. The chicks give a shrill and drawn-out trilling call when begging for food.

Habitat It inhabits dense deciduous or coniferous forest, with mature trees, including some that are damaged with cavities that are large enough for nest-sites. It also needs access to fish-rich water that does not freeze over completely in the winter – usually fast-flowing streams or rivers but also freshwater springs, and lakes in some areas. In the north it may fish on undisturbed coastlines. Although it has a reputation for being a very shy species, it is not entirely averse to breeding on the edges of human habitation, and in woodlands that are subject to some level of disturbance.

Behaviour, hunting and diet Though usually nocturnal and solitary (the pair bond notwithstanding), Blakiston's Fish Owl is occasionally observed fishing by day and in small groups, such behaviour no doubt dictated by changes in access to suitable fishing grounds during severe weather. Otherwise, activity normally begins at dusk. The day is spent roosting in a con-

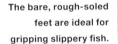

The bare, rough-soled feet are ideal for gripping slippery fish.

cealed spot in a tree. This owl spends much time on the ground, walking alongside rivers in search of prey or making its way between hunting perches.

It hunts by watching and waiting from a perch close to the water's surface – a low branch or the edge of an ice sheet. It may stand and watch for very long spells before making a strike, rather like a heron. When it spots prey it either drops straight down onto it or glides, catching the prey in shallow water or on the shore. It will normally carry small prey away to another perch to eat it, but larger items are partially eaten *in situ*.

Anything that moves in or near the water is fair game – fish of various sizes are taken, salmon and related species in particular, but also crayfish, frogs and waterbirds, along with the occasional small or medium-sized mammal. There is considerable seasonal variation in the make-up of the diet as different prey types fluctuate in availability – in spring frogs become a very important food source, but unsurprisingly they are more or less absent in winter, when more waterbirds may be taken. Fish form a key part of the diet at all times, hence the species' year-round dependence on accessible ice-free stretches of water.

Breeding The pair will normally stay together for successive years, defending the territory year-round. These large owls need large nests. Hollows in large mature trees are the usual choice but they will also use other birds' abandoned nests or nest among rocks or platforms created by forks between large branches or places where a branch has broken. There have also been reports of nests on the ground. They are willing to use suitably sized nestboxes, especially if these are constructed carefully to resemble natural cavities. Nesting is not attempted every year, but only when food supplies and weather are good.

Courtship calling begins in January. The female lays one or two eggs in early spring, incubating alone for about 35 days while the male brings food. Once the eggs hatch she will help with the hunting during the night but remain with the chicks through the daylight hours. The chicks take another 40 days to fledge and need several months of parental care before they are independent. However, they will remain in or near to their parents' territory until the following summer, apparently tolerated by the parents, as groups of several owls have been seen perched together in the same tree in early spring. Birds may pair up when just two years old, but most pairs do not attempt to nest until at least three years old.

Status and conservation This owl's world population is estimated at around 1,000 individuals. Of these, about 200 belong to the nominate island subspecies, most of them living on Hokkaido and Kunashir Island in the Kurils. The mainland population of some 800 birds is mainly located in south-east Russia. It is thought to be in steep decline.

Many threats face this species. Its riverine forest habitat is heavily logged, with suitable nesting trees (which are already very scarce) being removed. It is also affected by overfishing, habitat disturbance and damage from the creation of dams, accidental trapping by fishermen and direct hunting – in some areas it is killed for food, or because it preys on fish. Its slow breeding rate means its ability to recover its numbers is limited. It has been known to be preyed on by lynxes, which is probably a very rare occurrence but with such a small population any predation is potentially problematic.

In some areas conservation measures, including provision of nestboxes, are helping to address the problems – on Hokkaido, where people hold the owl in very high regard, protection measures include provision of supplementary food as well as nestboxes. More conservation measures are urgently needed elsewhere in its range, including a ban on fishing in some areas and strict protection of riverine forests. The Blakiston's Fish Owl Project is coordinating a number of projects intended to help improve the bird's status, including population surveys in the less well-studied parts of its range, and the development of conservation plans for key sites. Because of its small population, small range and declining numbers, it is assessed as Endangered.

A Blakiston's Fish Owl stares into the water, intently searching for prey movement.

Brown Fish Owl
Bubo zeylonensis

The Brown Fish Owl is a bird of dense riverine forest.

Size 48–56cm

Range This owl occurs in southern and south-eastern Asia, patchily in Turkey, then along the southern edge of Iraq across to the Indian subcontinent, and on to south-east China, Thailand, Laos, Cambodia, Vietnam and northern Malaysia. A separate small population in Israel is probably now extinct.

Evolution and relationships See Blakiston's Fish Owl (page 142) for discussion of the fish owls within the genus *Bubo*. DNA studies show that the Brown Fish Owl is most closely related to the Buffy Fish Owl, which replaces it further south through Malaysia. An extinct *Bubo* owl, *B. insularis*, which lived in the eastern Mediterranean region during the late Pleistocene era some 100,000 years ago and known from fossil remains, is considered by some authorities to be conspecific with the Brown Fish Owl.

Description This fish owl is a similar size to the Tawny Fish Owl, but in other respects looks similar to Blakiston's Fish Owl. The mantle feathers are light brown with a strong dark central streak and finer cross-barring, while the coverts are similar but with pale fringes. There is a subtle pale scapular line. The primaries, secondaries and tail feathers are boldly barred with dark and light brown. The underside is brown, marked with strong dark cross-barred streaks up to the chin and upper breast, which are contrastingly whitish and only finely streaked. The nape and crown are dark-streaked. The facial disk is only vaguely defined with no clear border, and marked with fine streaking, radiating outwards. It has large shaggy ear-tufts on the head corners, pointing outwards. The eyes are yellow, the bill and claws horn-coloured, and the tarsi and toes are bare. In flight it shows broad wings and strong banding on both wings and tail. Juveniles are a little duller than adults.

Geographical variation There are four subspecies. The nominate is found on Sri Lanka. On the mainland, the larger and paler form *B. z. leschenaultia* occurs in India and east into Thailand, replaced further east to Vietnam, China and Malaysia by the darker *B. z. orientalis*. The very pale subspecies *B. z. semenowi*, which may warrant full species status, is found in the more arid western part of the range, across the Middle East to Pakistan.

Right: This species shows the rather small head and unfeathered feet typical of fish owls.

Movements and migration It is a resident and sedentary species.

Voice It has a very deep, emphatic hooting song of three notes, the middle note longer and lower than the first and last. It repeats this two-second phrase every few seconds.

Habitat It is closely associated with fresh water, still and running, and inhabits mainly quite thick and dense lowland forest, sometimes more open woodland. The western subspecies *B. z. semenowi* occurs in areas that are generally drier and less vegetated but still sticks close to large rivers and other permanent water bodies.

Behaviour, hunting and diet The Brown Fish Owl is primarily nocturnal but may be seen hunting by day or bathing at the water's edge. When not actively hunting it rests in a large tree or rocky crevice. It hunts from a perch near the ground or sits on the ground or even in the middle of a stream, watching for prey and regularly flying or wading to a new spot. When it spots movement it pounces on the prey or swipes it out of the water. The diet is composed of a variety of aquatic animals, including fish but also frogs, crustaceans and insects, and sometimes carrion.

Breeding Pairs stay together in the same territory for successive years. The breeding season begins in November or December depending on location. It needs a sizeable platform or crevice of some kind for a nest-site – this may be a rocky ledge, a tree cavity or the old nest of an eagle or other large bird. It lays a clutch of one or two eggs which the female incubates for about 38 days. The chicks fledge after seven weeks or so.

Status and conservation This owl is vulnerable to disturbance, pollution and deforestation, and in the west of its range its numbers have certainly declined, with some breeding areas apparently abandoned altogether. In India, many are killed for superstitious reasons. However, elsewhere (in Sri Lanka, for example) it can be relatively common. Its population size is not currently known, but it has a large range and is not thought to be in severe decline across much of that range. It is therefore assessed as Least Concern.

Tawny Fish Owl
Bubo flavipes

A little-known species, the Tawny Fish Owl lives in remote upland woods.

Size 48–58cm

Range This owl is found in south-central and south-east Asia, from Afghanistan and across in a broadening band through northern India to south-east China, Thailand, Laos and Cambodia, and also Taiwan.

Evolution and relationships See Blakiston's Fish Owl (page 142) for discussion of where the fish owls, formerly placed in the genus *Ketupa,* fit within the genus *Bubo*. The precise taxonomic relationships of the Tawny Fish Owl have yet to be investigated.

Description This is a large, stocky but long-legged owl with a rather shaggy appearance. The upperparts are mostly dark with lighter brown barring and fringes. The breast, belly, crown and nape are a warm light brown with dark streaks that become narrower and longer further down. The facial disk is not clearly defined. The shaggy ear-tufts point outwards from the head corners and are often held flat, although they can be raised. The eyes are yellow, the bill and claws greyish, and the tarsi and toes are bare. In flight the broad wings look strongly banded. The juvenile's underparts look more spotted than streaked.

Geographical variation It is monotypic.

Movements and migration The Tawny Fish Owl is resident and sedentary.

Voice The male's song is a deep buzzing three-note hoot, which the female answers with a higher mewing call.

Habitat This species lives in dense forests with streams, up to an altitude of 2,450m.

Behaviour, hunting and diet It is most active from just before dusk and through the early night, though may be seen hunting in broad daylight when feeding chicks. It usually hunts alongside the streams that cut through its habitat, searching for prey on foot or from a low branch. It takes mainly fish, frogs and crayfish but will also catch small mammals, including rats and porcupines, and birds up to the size of partridges.

Breeding The breeding season begins in November or December, but the pair bond is probably maintained throughout the year, being renewed with courtship behaviour at the start of the season. The species nests in suitable hollows in trees, on the ground or in disused eagle nests, and lays a clutch of usually two eggs. Little is known about breeding behaviour, but it is likely to be similar to that of Blakiston's Fish Owl.

Status and conservation This species' population is not known, and it certainly faces threats in parts of its range. However, it has a large range and there is no evidence of serious decline, so it is assessed as Least Concern.

Himalayan Wood Owl
Strix nivicola

Size 35–40cm

Range This owl occurs through the Himalayas and then broadly across the southern half of China to the coast and Taiwan, and south into Laos and north Thailand.

Evolution and relationships See Tawny Owl (page 150) for discussion of *Strix*. This species is a recent split from the Tawny Owl, based on distribution and voice, but is not recognised by all authorities.

Description It is similar to a grey morph of the Tawny Owl. The upperparts are grey-brown, the mantle feathers dark-streaked, the coverts dark-streaked and white tipped, and the longer wing feathers and tail barred various shades of grey-brown. The underparts are grey with heavy cross-barred streaks, as are the crown and nape. It has a Tawny Owl-like brown facial disk, outlined in black and with indentations top and bottom, with a whitish eye surround. The eyes are black, the bill and claws grey, and the toes and tarsi are feathered. Juveniles are more barred.

Geographical variation Three subspecies are recognised – the nominate across most of the range, the paler *S. n. ma* in north-east China and Korea, and on Taiwan the darker *S. n. yamadae*.

Movements and migration It is thought to be resident and sedentary but may undertake altitudinal migration.

Voice The song has a similar tone and fluty character to that of the Tawny Owl but is different in structure, comprising two or three rather short, equal-length notes on the same pitch. Its other calls are not yet known.

Habitat Upland forests are home to this species. It uses coniferous and deciduous woodland up to 2,650m. Unlike the Tawny Owl it has a strong aversion to living in areas near human habitation.

Behaviour, hunting and diet It is nocturnal, and its hunting behaviour is probably similar to that of the Tawny Owl. The diet is composed of various small mammals, birds and invertebrates.

Breeding In common with other sedentary *Strix* species, it is thought to maintain its pair bond long term. It nests in tree-holes or rocky crevices, and nesting activity begins in winter. Other details of its breeding biology are not known but seem likely to be very similar to that of the Tawny Owl.

Status and conservation This enigmatic owl's population size is not yet known, but it seems common in some areas and has a large range. As a recent split from Tawny Owl, it has no conservation concern rating as yet.

Rather Tawny Owl-like, this species has heavily barred underparts in juvenile plumage.

Tawny Owl
Strix aluco

Right: By day, Tawny Owls roost in inconspicuous spots, but are often found and mobbed by small birds.

Well-placed nestboxes can encourage Tawny Owls to breed in town parks and gardens.

Size 36–40cm

Range This owl is found across most of Europe, including Great Britain (but not Ireland) and the south of Sweden, Norway and Finland. It extends south to the larger Mediterranean islands and into north-west Africa and the coasts of north Arabia, and east across central Russia as far as western Siberia. Separate smaller populations are found in central Asia from Uzbekistan into China, and in the Himalayas in northern Pakistan.

Evolution and relationships The genus *Strix*, or wood owls, belongs to the subfamily Striginae within the 'typical owl' family Strigidae. It is closely related to the small genera *Jubula* (Maned Owl), *Lophostrix* (Crested Owl) and *Pulsatrix* (the spectacled owls). The Tawny Owl's closest relative is the Ural Owl, the two species diverging from their common ancestor between 4 and 5 million years ago.

Description This is a medium-sized and very rotund-looking owl with a large head and short wings and tail. It occurs in rufous, grey and intermediate morphs. Grey morphs predominate where the species uses coniferous woodland, while desert populations are paler.

The mantle is brown with strong dark streaks and there is a prominent white scapular line. The median and greater coverts are brown with dark central streaks and white tips, forming two variably prominent wingbars. The primaries and secondaries have subtle dark and light barring, as do the tail feathers. The nape, crown and necksides are brown, with darker fine streaks and a broad whitish V shape on the forehead, extending down towards the bill. The underside is light brown marked with strong dark cross-barred streaks.

The facial disk appears to be in two distinct circular halves, with indentations at top and bottom. It has a fine dark border and is rather plain brown, with a whitish cross of feathers centred on the bill base and framing the inner edges of the eyes. The eyes are black, the bill horn-coloured and the claws grey. The toes and tarsi are fully feathered. In flight it looks very short- and broad-winged and top-heavy. Juveniles are densely barred.

Geographical variation As this is a variable species, determining whether a form represents a different subspecies is difficult, and further study (especially on DNA

Hume's Owl
Strix butleri

Size 30–34cm

Range This owl is found in coastal areas around the Arabian Peninsula. Its distribution in the interior is not known. It also occurs on the other side of the Red Sea in northern Egypt.

Evolution and relationships See Tawny Owl (page 150) for discussion of *Strix*. Hume's Owl was formerly considered to be a subspecies of Tawny Owl and was known as Hume's Tawny Owl (*S. aluco butleri*), but was split as a full species in the late 20th century. DNA analysis confirmed that the split is appropriate and identified the African Wood Owl *S. woodfordii* rather than the Tawny as Hume's Owl's closest relative.

Description This is a chunky, large-headed owl, a little smaller than the Tawny Owl. The upperparts are light brownish-grey patterned with black and white. The mantle and inner coverts have dark central streaks, and there is a whitish scapular line and obscure whitish wing-bars. The underparts are whitish-sandy with fine, slightly darker barring with cross-streaks. The flight and tail feathers are quite boldly banded with blackish, whitish and sandy-grey. The nape and crown are yellowish-grey barred blackish, and there is a dark vertical crown stripe to the bill base. The light sandy-brown facial disk has a narrow dark border, and there is a whitish crescent framing the inner edge of each eye. The eyes are orange, the bill and claws a dark horn colour. The tarsi and toe bases are feathered. It looks pale and shows strong wing banding in flight. Juveniles look like fluffier and more barred adults.

Geographical variation It is a monotypic species.

Movements and migration As far as is known, it is resident and sedentary.

Voice The male sings in phrases that comprise one long hoot followed by two shorter double hoots. The tone is fluty, higher pitched than the Tawny Owl and has an emphatic, almost whooping quality. The female has a similar song, and the two will duet together. It may give an accelerating series of deeper notes in alarm. Its other sounds are not well studied.

Habitat Hume's Owl is a bird of arid, rocky habitats, including desert, desert fringes, dry riversides and mountainsides. It is usually found near whatever water sources are around – streams, springs or semi-permanent pools. It has no particular aversion to human habitation and does not require the presence of trees in its habitat.

Behaviour, hunting and diet It is nocturnal, spending the day hidden in a rocky crevice or other hiding place and emerging at dusk to begin hunting. It usually watches for prey from a perch, catching it on the ground after a quick swoop, but may also chase flying prey and capture it in mid-air. All small animals are on the menu, from mice and gerbils, small birds and reptiles, to moths, scorpions and grasshoppers.

Breeding The pair bond is long-lasting, and the pair remain together through the year. Vocal activity picks up from January, and nesting begins in March or April. The nesting chamber is usually located on a ledge in a rocky crevice of some kind, perhaps also an accessible corner of a disused building. Very little is known about its breeding behaviour. From the few observed nests, the clutch is in the region of five eggs, and as with other owls the female incubates alone, though there is an observation of a male apparently 'covering' a clutch while the female briefly left the nest. The incubation period is about 35 days, and the chicks fledge after about 37 days. They may be seen moving about on ledges away from the nest itself a week or two prior to the first flight.

Status and conservation Hume's Owl is still a very little-known species, and making meaningful estimates of its population, as well as assessing its status, is problematic. However, it is proving to be more common and widespread than was thought in the last century. Its habitat is quite unproductive from a human perspective and so likely to be fairly secure, although the use of pesticides around oases could affect prey populations. It is rather frequently killed by cars in Israel, a country estimated to hold about 200 breeding pairs. Overall, the population is thought to be stable, and because the known range is quite large it is currently assessed as Least Concern.

Spotted Owl
Strix occidentalis

Youngsters retain a little down around the head even after the body plumage is fully developed.

Size 40–48cm

Range This owl is rather patchily distributed in western North America, reaching north along the coast into south-east Canada and south to California, and in a separate population from the west-central US states south into north and central Mexico.

Evolution and relationships See Tawny Owl (page 150) for placement of the genus *Strix*. There is some justification for splitting this species into two, as one subspecies is very distinct, and its population is well separated from that of the nominate form. The Spotted Owl is often regarded as a close cousin of the Barred Owl, but the two diverged from a common ancestor between 6 and 8 million years ago.

Description The Spotted Owl is a relatively small-headed and long-tailed wood owl, with a very complex plumage pattern. The upperparts have a mid-brown base colour, with liberal small white spots on the mantle and coverts, changing to whitish bars on the flight feathers. The individual underpart feathers are banded dark and light brown and have round white notches on the sides, producing a very mottled overall appearance. Its nape and crown are finely barred brown and whitish.

The facial disk has a strong dark outline and is indented at the top and bottom so looks like two circles. It is marked with dark and light concentric circles. The eyes are outlined with thick whitish feathers on their inner edges. It has black eyes, a horn-coloured or yellowish bill and grey claws. The tarsi are feathered, as are the toe bases. The wings look dappled in flight. Juveniles have a simpler and more barred pattern.

Geographical variation There are up to three subspecies, though one is of doubtful status and the other is a candidate for full species status. The nominate form is found in Nevada and central and south California, while from Canada to north California is a darker form, *S. o. caurina*, considered by some to be a dark morph rather than a subspecies. The southern and Mexican population is designated *S. o. lucida*, which is paler and more strongly spotted. It is sometimes split as a full species – Mountain Forest Owl *S. lucida* – but genetic study is necessary to clarify its taxonomic status.

Movements and migration Once established in a territory it is resident and sedentary, only young birds ever wandering any distance. In the central Sierra Nevadas, where it breeds at elevations of above 2,000m, it can be an altitudinal migrant, with winter movements down the slopes of up to 32km, an elevation drop of about 70m. Young birds in their first winter may wander considerable distances as they seek their own first territory, often travelling between 50 and 100km from their natal home.

Voice The territorial song comprises four quite powerful hoots, the middle two more closely spaced than the first and last. Both sexes use this call, and it appears to serve as an 'I am here' call between the pair as well as a territorial advertisement song. A more forceful version is given in agitation. Females dealing with a territorial challenge give a series of loud barking calls. The species' quite extensive vocabulary also includes assorted grating and yelping notes, and pairs exchange soft croons during mutual preening and other close-quarters courtship activity.

Habitat This owl is mainly found in very mature pine or mixed forest, with some trees of 200 years old or more. It prefers woodland with an uneven canopy height. It also requires plenty of shade and water. It will breed at both low and high elevations, though higher altitudes are used more in the south. The southern subspecies *S. o. lucida* will use more open areas, including shady canyons and partly wooded mountain slopes.

Behaviour, hunting and diet It is a nocturnal bird, and will roost in a shady spot in a large tree, either among the foliage or in a hole. It may also roost among rocks. It is noted for being extremely approachable when roosting. Sometimes it will be active by day though not necessarily hunting – it may, for example, bathe or visit a food cache. Usually it becomes active within half an hour after sunset and returns to its roost just before dawn. It usually hunts from a perch, swooping down on its target, and much of its prey is caught among the mid-level branches rather than on the ground. If the initial swoop is unsuccessful, the owl may chase after its fleeing prey on foot.

Its diet is composed mainly of small mammals (up to 90 per cent by weight) and birds. In some areas it seems particularly dependent on flying squirrels and wood rats – studies have shown that breeding pairs that take mainly large prey items like these have more success than those that take smaller prey on average. Less frequently taken prey includes hares, bats and assorted invertebrates.

Breeding Pair bonds in this species are long-lived and maintained all year. The pair may roost together outside the breeding season, but in general live fairly separate lives through autumn and early winter. They begin to call more and spend more time together and near their nest-site in late winter. Up to 12 days before she lays her first egg the female spends time in the nest, which is either a tree cavity or an old stick nest of another bird, with the former apparently being more widely used.

The male starts to feed the female when she begins to spend time in the nest, and the pair copulate regularly, usually just after sunset. In due course a clutch of

The Spotted Owl is noted for being unusually 'tame' and easy to approach.

two to four eggs is laid, the female beginning to incubate as soon as she has laid her first. The eggs hatch asynchronously after about a month, and the female remains with the chicks for another couple of weeks while the male brings food, which she tears up and feeds to the small chicks piece by piece. When the young birds are about two weeks old, the female begins to hunt for the brood along with the male, and both parents must work hard to keep the chicks alive – mortality in the nest due to starvation is rather high. Both parents bring back prey that has been decapitated, apparently deriving enough nutrition from the heads to sustain themselves.

The chicks venture out onto nearby branches at about 35 days old. At this age they frequently fall to the ground while testing their growing wings but show strength and determination, if not finesse, in climbing back up into the branches. However, some owlets do not quickly make it back to the safety of the trees and may run the gauntlet of ground predators for several days. After a week or two they are able to fly. They are independent when aged about two and a half months and disperse in search of their own territory. This is a difficult time, as established birds occupy the best patches, and youngsters may wander some distance to find even a marginally suitable winter territory. There is some evidence from radio-tracking studies that the shorter the distance a young Spotted Owl disperses

from its natal grounds, the better its chances of survival to the following spring. They will generally make their first breeding attempt when two years old.

Status and conservation This owl is declining rapidly, mainly due to intensive logging in its habitat. Even selective logging can change forests too much for the species' tolerance, as it requires dense forest. Competition and hybridisation with the Barred Owl as that species, which is larger and more adaptable, spreads west is another threat. There is great concern about the Spotted Owl's future, both globally and in each of the states where it occurs. It is close to extinction in British Columbia, its only breeding area in Canada, where a captive-breeding programme is under way to try to save it. In the same state, a cull of Barred Owls is under consideration.

Protecting the species is problematic for several reasons. In order to save it, radical reform of the logging industry is needed to save the old-growth forests, but this will threaten jobs in an industry that has already suffered due to previous reforms and because timber stocks are on the wane. There is also controversy surrounding the idea of culling Barred Owls. As things stand, the Spotted Owl population is estimated at just 15,000 birds – this low population and the rate of decline means it is assessed as Near Threatened.

Above left: Forest clearance and competition with encroaching Barred Owls threaten this species.

Above right: 'Brancher' owlets move from the nest to nearby perches at about five weeks old.

Barred Owl
Strix varia

The barred upper breast helps separate this species from the Spotted Owl.

Size 48–55cm

Range This owl is found in North America from southern Canada down to Mexico. The main part of its range is the eastern half of the continent, but it is also found in a narrow band across to the north-west and then spreads south from there. It generally replaces the Spotted Owl but the two occur together in the north-east, and the Barred Owl is spreading south into that species' core range.

Evolution and relationships See Tawny Owl (page 150) for the placement of the genus *Strix*. This species is apparently closely related to the little-studied Rufous-legged Owl *Strix rufipes* of southern South America. Although it diverged from its common ancestor with the Spotted Owl at least 6 million years ago, the two species are still able to interbreed and do so readily where their ranges meet, which is a problem for the threatened Spotted Owl.

Description This is a fairly large, wide-headed, stocky owl with strongly patterned plumage. The upperparts are mid-grey-brown, and the mantle and inner covert feathers have bold white spots on the tips. The outer coverts

Right: This species is one of North America's commoner owls and is expanding its range.

and flight feathers are banded brown and whitish, as are the tail feathers. The underparts have a very light brown ground colour, and strong dark barring on the breast, thickening further down and then transitioning abruptly to broad dark vertical streaks on the belly. The crown and nape are light brown, with darker bars.

The facial disk is wide and indented top and bottom, forming two round halves. The disk has a fine blackish edge, and inside is whitish marked with concentric brown circles. The eyes and bill are framed with a central cross of whitish feathers. The dark brown (black from any distance) eyes look rather small and close-set. The bill is a bright yellowish horn colour, the claws greyish. The tarsi are feathered but the toes are only lightly feathered or bare. In flight the strong wing-barring is striking. Juveniles are barred rather than streaked on the belly.

Geographical variation There are four subspecies of Barred Owl. The nominate occupies the bulk of the range, from south-east Alaska and south-west Canada down to northern California, east across southern Canada to the Atlantic coast, extending south as far as Oklahoma and south-east to Virginia. South-east of here it is replaced by the paler and more yellow-toned

Most prey is caught on the ground, but the Barred Owl is a versatile hunter and uses many techniques.

S. v. georgica. The light brown form *S. v. helveola* occupies north-east Mexico, while in southern Medico is the very dark *S. v. sartorii*.

Movements and migration It is resident and sedentary at most times but will wander when necessary to find feeding grounds. Ringing studies show that among adults at least, movements of more than 10km are very uncommon. Youngsters, obliged to wander to find their own territories, travel further. One bird ringed in north-east Canada moved 1,600km west to Ontario.

Voice The male's usual song is a series of emphatic short hoots, rising in pitch and ending with a staccato chuckle that falls back down the scale. The tone is rather squeaky, lacking the resonant tones of most other *Strix* species' calls. The female responds with a similar song but hers is higher-pitched and even squeakier. The male also has a slower-paced version of the song. Pairs duetting often sound excitable, with a frantic, almost gasping quality to the notes. Young birds beg with a shrill, grating squeal, while the female's begging call to the male is similar but lower-pitched.

Habitat The Barred Owl inhabits mature forest, both low-lying and in the uplands, often in wet and swampy

areas or with lakes around. It will use quite dense forest but prefers more open woodlands with glades and the edges of forest where it meets rough open ground. It will also breed in large parks and suburban areas with good-sized gardens, provided there are stands of mature trees. Most birds use deciduous or mixed woodland, ideally with a reasonable number of old and damaged trees for nesting purposes.

Behaviour, hunting and diet This is a nocturnal species, hiding in a roost among foliage high in a tall tree. Songbirds will mob the Barred Owl if they discover it, which leaves it exposed to the risk of attack by other birds of prey, so effective concealment is a necessity. It will hunt by day when needs must – at the height of the chick-feeding period and when food is sparse in winter.

Its preferred hunting method is to sit and wait on a favourite perch, swooping down on prey when it sees or, more likely, hears it below. It will also chase and catch flying prey on the wing and will attempt to flush birds from their roosts by flying close to hedges and bushes and striking the twigs with its wing-tips.

The diet is very diverse. As with the other wood owls, small rodents such as voles, mice and shrews form the core of the prey taken – between 50 and 75 per cent, but other mammals on the menu include bats, mustelids, rabbits, hares, opossums and squirrels. It readily tackles birds up to the size of crows and pheasants, and will hunt smaller owls, including the very feisty Eastern Screech Owl and the Long-eared Owl. It may wade in shallow water to hunt frogs, fish and in some areas even baby alligators, and takes a small number of invertebrates.

Breeding This species usually pairs for life, and male and female stay together year round, defending their territory. Particularly good nest-sites may host nesting Barred Owls for decades, with pair members quickly replaced when they die. Unpaired birds sing from autumn, but established pairs are most vocally active from winter. Their courtship peaks with much very excited vocalisation, and the male may dance along a branch and flick his wings at the female before they copulate. They will also preen each other in more sedate moments, and courtship feeding takes place, becoming more frequent as the female gets ready to lay her eggs.

Nest-sites vary from the old nests of other birds or old squirrel dreys, to tree cavities. The female begins to lay in March and produces two or three eggs, exceptionally four or five. Her incubation begins immediately the first egg is laid, and lasts between 28 and 33 days (longer in the north of the range). She broods and feeds the chicks for three weeks after hatching, tearing prey up into small pieces at first, and then giving larger morsels with some fur and bone after the first week or so, whereupon the chicks begin to produce pellets. As the chicks grow, they become noisier and more demanding, and are soon strong enough to stand and scramble around. They leave the nest at about four weeks old and climb into nearby branches where they sit and await food deliveries from both parents, revealing their presence with their grating begging calls. The parents are very protective of the nesting area at this time and will readily attack anyone who wanders too close.

By the age of six weeks the young owls are making their first tentative flights and have shed most of their nestling down. Over the next two months they will begin to hunt for themselves, beginning with easy quarry like invertebrates before moving on to small mammals. Parental food deliveries become less frequent, ending altogether when the chicks are four months old. In some parts of the range the pair may then attempt a second brood if food is abundant.

Status and conservation The Barred Owl is a fairly common species over much of its range, and its population is estimated to number a healthy 600,000 or so birds. It is increasing and expanding its range, its spread apparently assisted by logging in the north and west as this creates a more open forest structure, although of course complete forest clearance will force it to abandon an area. The use of nestboxes to replace natural cavities in felled old trees helps it to maintain a population in affected areas.

The impact of the Barred Owl on its Near Threatened relative the Spotted Owl is currently subject to much investigation, and measures to discourage Barred Owls from occupying Spotted Owl habitat are being explored. The Barred Owl is larger than the Spotted and will outcompete it in disputes over nest-sites; the two also hybridise. The Barred Owl itself is not of conservation concern because of its wide range and large, increasing population; it is evaluated as Least Concern.

A well-grown chick is brought quite large prey items to handle alone.

Ural Owl
Strix uralensis

The Ural Owl is one of the suite of species adapted to live in Eurasia's boreal forests.

Size 51–61cm

Range This owl is found from eastern Scandinavia in a broadening band east across north-central Russia to the coast, across to Sakhalin island and Japan, and south into eastern China. There are also some small isolated populations in Poland, the Czech Republic, Slovakia, Slovenia, Croatia and the Balkans. It has been reintroduced to Germany.

Evolution and relationships See Tawny Owl (page 150) for discussion of the genus *Strix*. The Ural Owl is closely related to the Tawny Owl. A population in the Sichuan Mountains in China has recently been split by some authorities as Sichuan Wood Owl *Strix davidi*.

Right: It is similar to the Tawny Owl in appearance, but larger and longer-tailed.

Description This is a large owl, overlapping in size and weight with the Great Grey Owl. It is large-headed and rather long-tailed. The general impression is of a rather ghostly light grey bird, though there is subspecific variation, and in some areas a brown morph occurs. Its mantle is light grey, each feather marked with a strong dark streak. There is an obscure white scapular line. The coverts are grey with dark central streaks and white tips to each feather, changing to banding on the greater coverts and strong dark grey and whitish banding on the flight feathers and tail. The head, nape and underparts are light grey with strong but fine dark streaking, longest and finest on the belly.

The large facial disk is clearly outlined greyish, and is indented at the top and bottom, so it forms two round halves. Inside, the disk is light grey with a whitish cross framing the inner sides of the eyes. The eyes are dark brown (looking black at any distance) and appear rather small within the large facial disk, giving the bird a mild expression. The bill is quite a bright yellowish horn colour, striking in its grey face, the claws grey. The toes and tarsi are feathered. In flight it looks powerful, long-

tailed, heavy-headed and with prominent banding on the wings. It lacks the Great Grey Owl's pale patch at the primary bases. Juveniles are barred rather than streaked on the underparts.

Geographical variation Some eight or nine subspecies are recognised. The nominate, which is the palest subspecies, is found in the east of the range including Siberia and the southern Urals. The form *S. u. liturata*, darker than the nominate, occurs in Scandinavia and the Baltic countries, intergrading with the nominate around the Volga, its eastern limit. Three different subspecies occur in Japan – *S. u. fuscescens* in the south, the larger *S. u. hondoensis* in central parts and the small *S. u. japonica* on Hokkaido in the north – all are more reddish-toned than the nominate. In central Europe, the isolated populations belong to the large, dark and strongly marked form *S. u. macroura*, which occurs in a brown morph as well as grey. On Sakhalin and in north-east China and Korea is *S. u. yenisseensis*, a dark and heavily marked form.

Both members of the pair are watchful and very defensive of the nest-site.

Movements and migration It is resident and sedentary, but northern birds in particular may undertake cold-weather movements, and juveniles wander up to 150km in search of their first territory.

Voice Territorial males have a hooting song, consisting of a longer note followed by a short stuttering sequence, with gaps of up to 50 seconds between the phrases. The hoot is quite deep, booming and resonant, and lacks the sweet, quavering and fluty tones of the Tawny Owl's hoots. The female responds with a similar song but hers has a higher, more yapping tone. It has a Tawny Owl-like but deeper 'ku-vit' contact call and an abrupt barking alarm call.

Habitat This is an owl of mature pine and mixed forest, with glades and other open areas. It is usually found in the lowlands and sometimes in quite boggy woodlands, but in the southern parts of its range it will use more hilly forest, up to 1,600m in parts of central Europe. It will live alongside human habitation provided the surrounding habitat is right, visiting places where spilled grain attracts small birds and rodents, and may even live on the fringes of cities. The central European population mainly inhabits beech woodland.

Behaviour, hunting and diet A nocturnal species, the Ural Owl will also hunt by day when driven to by the need to provide for well-grown chicks or in winter when food is scarce. Normally it roosts by day high in a tree, concealed by foliage. Activity begins around dusk. It

hunts mainly from a perch, but may also use a quartering flight, especially over more open areas. Like the Great Grey Owl, it is capable of hearing the movements of rodents under a substantial snow layer and will accurately strike through the snow to catch prey.

It hunts mainly voles (they form up to 90 per cent of its diet), and its breeding success is highly influenced by vole population fluctuations. Other rodents such as lemmings are also taken. Birds make up a small proportion of the diet but a wide range of species may be taken; it is certainly capable of handling prey the size of Woodpigeons or grouse. At the other end of the scale, it will also take insects and frogs at times, and will eat carrion (this may be particularly important to 'homeless' territory-seeking juveniles in winter).

Breeding The pair usually remains together for life, and together they defend a large territory all year round. Unpaired birds will sing from autumn, drawing maximum attention from potential mates by moving around their territory and singing in flight as well as from a selection of favourite perches, but the main courtship period begins in January. The pair duet together, and the male begins to feed the female and to show her possible nesting places if they do not already have a previously used site. The nest-site varies according to what is available – it may be a cavity in a decaying tree, a partial hollow left in a trunk where a branch has fallen, the old stick nest of a large bird, such as a buzzard or Goshawk, a crevice among rocks or even a sheltered hollow on the ground. It will also use nestboxes. In poor vole years pairs may not attempt to breed, although they remain on territory and guard against intruders.

The female begins to spend time in the nest about 10 days before she lays her first egg, in early spring. The male's food deliveries help her reach egg-laying condition. Usually three or four eggs are laid, at two-day intervals, but in good vole years the clutch may be up to six, with incubation beginning when the first egg is laid. The incubation period seems quite variable: 28–35 days, depending on latitude and perhaps other climatic factors. The male continues to hunt throughout this period and during most of the chicks' time in the nest, while the female remains on the nest with the chicks. Both parents will fiercely defend the nest from human intruders, diving and striking at the head with enough force to knock a person over and cause serious injury.

The young birds leave the nest at 35 days old and move into nearby branches. They are not fully feathered or capable of flight at this time, but they are strong on their feet and able to climb up to safety if they fall to the forest floor – they are also closely guarded by their parents during this vulnerable time. Over the next week or two they become stronger and practise beating their growing wings, making their first flights at 45 days old. They will continue to beg for food from their parents for another month at least but are independent at around three months old. They must then seek their own territories, and while they become sexually mature the following year are unlikely to breed until their second full year of life.

Status and conservation The Ural Owl is a very widespread species, occupying vast tracts of continuous habitat across Asia. It is vulnerable to deforestation, and, on a local level, to the loss of nest-sites where old and damaged trees are removed from the forest. In Finland and Sweden, placing nestboxes in suitable habitat has allowed substantial spread, increases in numbers and improved nesting success. It has a population of somewhere between 500,000 and 8 million birds, and the general population seems to be stable; for these reasons it is assessed as Least Concern.

Above left: Owlets venture from the nest when still small and fluffy.

Above right: Fledged chicks can climb strongly even though they are not yet able to fly.

A Great Grey Owl can accurately target prey moving under a layer of snow.

Great Grey Owl
Strix nebulosa

Length 65–70cm

Range This owl can be found across nearly the whole northern hemisphere, from Sweden and Finland across Russia (and edging into north Mongolia and north-east China). Its distribution continues across Alaska, Canada and the northern US states.

Evolution and relationships See Tawny Owl (page 150) for discussion of the wood-owl genus *Strix*. This species is closely allied to the Tawny and Ural Owls, though not as closely related as they are to each other.

Description The Great Grey Owl is one of the largest owl species but lacks the bulk of the eagle owls – its dense plumage conceals a relatively slight frame. It looks quite long-bodied, with a proportionately very large head. The upperpart plumage is intricately patterned in various shades of grey, with cross-barred streaks on the mantle and inner coverts, a dappled white scapular line and alternate dark and light bars on the larger coverts, flight feathers and tail. The underparts are marked with thick, cross-barred dark streaks. The nape and crown are finely barred.

The very large pale grey facial disk is split at the bottom and is strongly marked with dark grey concentric circles, while the proportionately small eyes are bordered on the inside with large white crescents – this gives the face a rather intense and thoughtful expression. It has a white collar or moustache, broken by a black oblong patch directly below the bill, in the gap between the two halves of the facial disk, which has given rise to the German name for the species – *bart kauz* or 'bearded owl'. The eyes are yellow, the bill horn-coloured and the claws greyish. The tarsi and toes are fully feathered.

In flight, the owl looks very front-heavy and quite long-tailed. It has the typical short, broad wings of the *Strix* owls, and shows a pale patch at the base of the outermost primaries. It flies with a stately, almost heron-

Its very top-heavy
appearance and unique
face pattern help
identify this owl.

like grace. Young birds have a smaller and less well-defined facial disk.

Geographical variation

Two subspecies are recognised. North American birds are of the nominate subspecies, larger and darker Eurasian birds belong to the subspecies *S. n. lapponica*. Behavioural observations and genetic studies indicate that the birds of the Sierra Nevada area of California may constitute a third subspecies, provisionally named *S. n. yosemitensis*.

Movements and migration

This is not so much a migrant as a nomadic wanderer, moving unpredictably. The most northerly populations of Great Grey Owls are at the mercy of the weather and, more importantly, the availability of their limited range of prey. A shortage of voles will force the birds to roam widely, most often in a southerly direction. Radio-tracked birds have been shown to move up to 650km over three months. In times of serious food shortages, there can be irruptions involving numerous birds moving out of their usual range in search of more productive hunting grounds. In the winter of 1980–81, hundreds of Great Grey Owls from Russia moved south and west into Finland and beyond, and a similar irruption took place in North America in the winter of 2004–05. The latter saw more than 5,000 Great Greys arrive in the state of Minnesota, over 13 times more than the previous state record. Some birds involved in these irruptions may stay in their new location and breed there, if supplies of voles remain high.

The Great Grey Owl often uses the old nest of another large bird.

Voice

The male's song is a deep but mellow-toned, rather pumping single hoot repeated 10 or 12 times in quite quick succession (one every 0.5 seconds or so), and accelerating towards the end of the sequence. The female has a similar song but her notes are harsher toned and higher pitched – she may duet with the male but will also sing when not yet paired up. Her begging call in courtship is a rather high-pitched up-slurred note. When pairs preen one another they give a soft crooning call.

Habitat

The Great Grey Owl nests in forests, which in the latitudes it inhabits are usually composed of pine and fir trees, sometimes also birch. In North America it makes use of poplar stands, and it will also breed close to the tree line in stunted forest. However, it prefers to hunt over fields, moorland or boggy open ground, and so the best areas tend to be forest edges or forests with good-sized clearings. Lowland areas are best, but it may also live in hilly forests.

Behaviour, hunting and diet

This owl is most likely to be active around dawn and dusk but will also hunt in the night (especially in winter) and sometimes even in the middle of the day (especially in summer). It roosts in well-hidden spots in trees, often pressed against the trunk for camouflage. If disturbed it adopts a tall posture, and the head shape changes, the top of the head becoming flatter and squarer.

It sometimes employs a 'sit and wait' hunting tactic, using a post or other vantage point from which to look

or listen, but also hunts actively from a low, quartering flight. Its preferred hunting periods are early morning and late afternoon, but it will also go in search of prey at any other time when necessary. It may drop onto prey steeply or from a long shallow glide. Thanks to its acute hearing it can detect its prey through a thick covering of snow. The attack has enough force to punch through a significantly deep snow layer – 30cm or more.

Despite its large size, this owl is primarily a hunter of small rodents, especially voles and, in North America, pocket gophers. It will also take small birds. At the upper end of the prey sizes it will tackle are rabbits and grouse, but vole-sized rodents generally form 80–90 per cent of the diet, with the highest proportion in the summer months. This heavy dependence on vole numbers dictates much of the species' general ecology, including aspects of breeding behaviour and dispersal patterns.

Breeding The pair bond is not necessarily renewed in successive years, as birds are quite inclined to roam and establish new territories depending on food supplies. However, there are records of a pair using the same site for at least five successive years. Males usually begin to give the territorial song in late winter, intensifying in early spring. Courtship is a protracted affair, with much courtship feeding and mutual preening, and inspection of at least one and sometimes several possible nest-sites.

This species usually nests in an old stick nest of another bird, usually a crow or large raptor species. It may also nest in a high broken-off tree stump. The female begins to lay in late March or April, and her clutch is usually of three to six eggs but in very good

prey years as many as nine. However, in very poor years, pairs may not attempt to nest at all. She begins to incubate the clutch as soon as the first egg is laid, and lays successive eggs at one- to three-day intervals. The incubation period is 28–30 days. During this time the male brings food to the nest for her.

When the chicks hatch, the male ups his hunting rate, and the female carefully tears up the prey into small pieces and feeds them to the chicks. As they grow, they can cope with larger and then intact prey, and begin to produce pellets when they are a week or two old. They leave the nest before they can fly, at three or four weeks old, and climb into nearby branches where they wait for their parents to bring food. Should they fall from the tree at this stage, they are capable of climbing back up into the branches, though of course are at risk from ground predators if they are not quick enough. The ability to fly comes gradually over the next four weeks, and even then they are still dependent on care from their parents for another two months at least. They become sexually mature the following year.

Status and conservation In Eurasia, this bird's dependence on forests in the pine zone makes it vulnerable to habitat loss from deforestation, but North American populations seem more willing to accept a variety of habitat types. Overall, the Great Grey Owl has a world population of around 60,000 birds. The population in North America has increased substantially since 1970, but overall numbers seem to be generally stable, except for fluctuations in response to vole numbers. Populations of voles tend to be cyclic, with one 'boom year' in every three or four. Its conservation assessment is Least Concern.

The Great Grey Owl is most active at twilight, hunting from flight or a perch.

Brown Wood Owl
Strix leptogrammica

Above left: This owl roosts high in the trees of its forest habitat by day.

Above right: A young bird with fully developed flight feathers but still much down on the body.

The dark surrounds to the eyes make them look even larger.

Size 40cm

Range This owl is patchily distributed in the Indian subcontinent and south-east Asia. It ranges to the west coast of India and Sri Lanka, discontinuously north to north-east India and Bangladesh, then south through the Malay Peninsula to Sumatra and Borneo.

Evolution and relationships See Tawny Owl (page 150) for discussion of *Strix*. This species' taxonomy needs further study but some former subspecies have now been split as full species.

Description A distinctive bird, it has mid-brown upperparts, darkest on the back, with darker barring on all feathers, most widely spaced on the flight and tail feathers. The breast is dark brown, grading sharply to lighter brown on the belly with fine dark bars. It has a whitish chin, dark nape and crown, and a pale brown facial disk with large blackish patches around the eyes, and short white eyebrows. The eyes are dark brown, the bill and claws a greyish horn colour. The tarsi and toes are lightly feathered. In flight it looks top-heavy and short-winged. Juveniles have clearer barring.

Geographical variation Of the four or five subspecies, the nominate occurs on central and southern Borneo, the darker *S. l. maingayi* in Thailand, Burma and the Malay Peninsula, while separately in India the similar *S. l. indranee* occurs.

Movements and migration It is resident and sedentary.

Voice The song has a typical fluty *Strix* tone, and begins with a single longer hoot followed by a series of shorter stuttered notes. It lasts about a second and is repeated every few seconds. It has various other shrill and growling calls.

Habitat It inhabits dense lowland forest.

Behaviour, hunting and diet It roosts by day on a high and well-hidden perch among foliage, becoming active after dark, and watching for prey from a perch. It usually catches prey on the ground. The diet is varied, with most small animals in its habitat represented, including mice, birds, reptiles and occasionally fish.

Breeding Nesting may begin as early as January, following a courtship period in which the birds call constantly on clear nights. It is thought to maintain lasting pair bonds. It nests in tree-holes or more rarely in open spots, such as cliff ledges or in the crook of a large branch. The clutch is usually of two eggs. Incubation and fledging periods are not yet known.

Status and conservation It is declining, and its habitat is threatened by deforestation. However, because of its large range it is still currently assessed as Least Concern.

Collared Owlet
Glaucidium brodiei

Size 15–17cm

Range This species occurs throughout south-east Asia, including Taiwan and parts of Java and Borneo, and extends narrowly west to north India and into Afghanistan.

Evolution and relationships The genus *Glaucidium* is placed in the owl subfamily Surniinae. The Collared Owlet's precise relationships have yet to be studied.

Description This is a very petite but rather long-tailed pygmy owl, Asia's smallest owl species. The upperpart plumage and the flanks and breast-sides are dark grey-brown with fine whitish bars. The centre of the breast and belly is white, apart from a solid dark collar. The chin is white, the crown and nape brown with whitish speckles – there are imitation 'eye-spots' on the nape. The dark facial disk is marked with subtle concentric pale circles, and it has narrow white eyebrows. The eyes are yellow, the bill horn-coloured and the claws grey. The tarsi and tops of the toes are feathered. In flight it looks long-tailed and has rapid wingbeats. The juvenile has a streakier head.

Geographical variation The nominate occurs on the mainland, while Taiwan, Borneo and Sumatra each have a separate subspecies, the most distinct of which is *G. b. pardalotum* of Taiwan – a more rufous form with teardrop-shaped spots on the belly.

Movements and migration It is resident and sedentary.

Voice Territorial males have a whistling song, of four notes, in a 1...2-3...4 rhythm, repeated every few seconds.

Habitat It inhabits light evergreen forest and forest edges in the uplands, reaching altitudes of close to 3,000m in some areas.

Behaviour, hunting and diet It is active at dawn and dusk and sometimes by day and night too. It watches for prey from a high bare branch and comes down in a gliding swoop to catch its victims. Its diet is mainly birds, which may be up to its own size or even larger – it is fiercely mobbed by all kinds of small forest birds when they discover it roosting. It will also take small mammals and invertebrates.

Breeding Little is known about its breeding behaviour, but the pair bond is likely to last only the duration of the breeding season, which begins in March or April. It nests in woodpecker holes or other tree hollows and lays a clutch of three to five eggs.

Status and conservation Although its population is not known, it is considered common over much of its range. Deforestation is the main threat it faces, but it is presently evaluated as Least Concern.

A widespread owl, the Collared Owlet is not difficult to see in suitable habitat.

This is a very small species, but a powerful predator for its size.

Northern and Eurasian Pygmy Owls
Glaucidium californicum and *G. passerinum*

A Eurasian Pygmy Owl shows the pseudo ear-tufts that appear when it raises its crown plumage.

Far right: The highest point on a pine tree is a perfect look-out post when the Eurasian Eagle Owl is hunting.

The Eurasian Pygmy Owl is agile in the air with a fast and manoeuvrable flight.

Size 16–19cm

Range The Eurasian Pygmy Owl is found patchily across central and northern Europe, as far west as France and south into northern Greece, and then in a band across central Asia through Kazakhstan, Mongolia, northern China and southern Russia to Sakhalin island. In the New World it is replaced by the Northern Pygmy Owl, which is found in the western US and Canadian states and into Mexico.

Evolution and relationships These two species, among the most widespread in the genus *Glaucidium*, were until recently lumped as a single species. However, recent DNA evidence supports a split into two, and in fact indicates that all of the Old World pygmy owls form a distinct lineage from the New World species. As the Northern and Eurasian Pygmy Owls are very similar in both plumage and ecology, they are here discussed together.

Description The Northern Pygmy Owl is a small and rather long-tailed species that occurs in brown, grey and red morphs. The dark brown or red-brown upperpart plumage is marked with whitish bars on the flight and tail feathers, and white tips on the coverts and lower mantle feathers. The underparts are white with well-spaced dark streaks on the flanks, coalescing into solid dark plumage with fine white bars on the upper breast-sides. The crown and nape are dark with fine white speckles, and on the base of the nape are blackish, white-outlined false-eye markings. There are very small ear-tufts on the outer corners of the face, often not visible. The facial disk is greyish, and there are narrow white eyebrows. The eyes are yellow, the bill horn-coloured and the claws grey. The tarsi are feathered, but the toes are bare. In flight it looks rather long-tailed and has rapid wingbeats.

The Eurasian Pygmy Owl is very similar in appearance. Rufous morphs are less frequent in this species. The white upperpart barring is finer, and the underpart streaks are also finer and more uniform across the entire breast and belly. The nape eye-markings are bolder, and the toes have some feathering on the uppersides. Juveniles of both species have less contrasting markings.

Geographical variation The Northern Pygmy Owl is divided into three or more subspecies, though the existence of different colour morphs rather confuse the picture and the validity of the different forms is uncertain. The nominate is the most northerly form. In the southern states is *G. c. pinicola*, a more strongly marked form, while *G. c. swarthi* is a very dark form that occurs on Vancouver Island.

The Eurasian Pygmy Owl is generally considered to comprise two subspecies, the nominate in the western part of the range and the paler and greyer *G. p. orientale* from eastern Siberia to Sakhalin island and northern China.

Ferruginous Pygmy Owl
Glaucidium brasilianum

Size 17–20cm

Range This owl is found across much of the forested interior of South America, and north through Central America into Mexico and just into the south-western US states.

Evolution and relationships See Collared Owlet (page 173) for placement of the genus *Glaucidium*. This species has a complex taxonomy and may represent a superspecies. Some of its subspecies are split as full species by some authorities; further study is needed to fully clarify the situation.

Description This is a small but long-tailed pygmy owl, which shows considerable variation in colour across its range and also occurs in red, brown and grey colour morphs in the same areas. The nominate is described here. The upperparts are grey-brown, with whitish barring across the coverts, flight feathers and tail, and a white scapular line. The underparts are whitish, marked with soft elongated teardrop-shaped streaks, heaviest on the flanks and breast-sides and absent from the centre. The nape and crown are dark brown with short whitish streaks, and there are black, white-outlined false eye-spots on the base of the nape. The facial disk is barely defined. There are prominent but narrow pale eyebrows. The red or rufous morph is similar but with a strong reddish suffusion to the entire plumage, somewhat obscuring the white markings. The eyes are yellow, the bill horn-coloured and the claws grey. The tarsi are feathered, and the tops of the toes have thin feathering. In flight the long tail is striking, giving a hawk-like outline. Juveniles are plainer.

Geographical variation The nominate is found through most of Brazil, eastern Paraguay, north-eastern Argentina and north Uruguay. The form *G. b. ridgwayi*, sometimes split as Ridgway's Pygmy Owl, is found in north-west Colombia, Central America, Mexico and the southern US, and is variable but more strongly marked than the nominate, with dark, rather than pale, barring on the tail feathers. The subspecies *G. b. stranecki*, found in central and eastern Argentina and south Uruguay, is a larger and greyer form. Up to six other subspecies are recognised.

A red morph Ferruginous Pygmy Owl has vivid, almost ginger plumage.

Movements and migration It is thought to be resident and sedentary throughout its range.

Voice The male's advertising song is a long (up to 60) series of sweet, ringing and breathy whistles, of even length and spacing, but becoming a shrill, rapid twitter when the bird is excited. A common call is a series of rather insect-like chirps. The subspecies *ridgwayi* has a slower and more hollow-toned song, and a more yelping quality to its call.

Habitat This is mainly a bird of lowland open forest, scrubby areas, woodland edges and, in some areas, well-vegetated semi-desert.

Behaviour, hunting and diet It is crepuscular and partly diurnal but may be active at any time of day. In typical pygmy-owl fashion it keeps a watch for prey from a high and sometimes quite exposed perch, though it regularly attracts parties of mobbing songbirds. When excited it bobs its head and flicks its tail upwards.

It is a versatile predator but is most likely to take birds, including those considerably larger than itself – it has, for example, been recorded taking domestic chickens and forest doves. It also takes some small mammals and invertebrates. For its size it has large and powerful feet and claws, enabling it to 'lock on' to a larger bird and cause serious damage; it also bites repeatedly at its victim's head.

Breeding This species is solitary most of the year, individual birds defending their own territories, and previous pair bonds are not necessarily renewed. Males begin to call to attract a mate in late winter, singing from a perch and frequently visiting and calling from the vicinity of possible nest-sites. The usual nest-site is a tree-hole, either natural or an old woodpecker nest. The female lays eggs at two-day intervals, with a final clutch size of two to five, and begins to incubate when the last egg is laid. The incubation period is about 27–30 days, during which the male brings food for the female.

The owlets are brooded continuously for the first week or two, and then left alone as both parents hunt for them. They leave the nest at about four weeks old but require another two or three weeks' care before they are independent. They are ready to breed the following year.

Status and conservation Habitat loss due to deforestation threatens this species in the most heavily forested parts of its range, while habitat degradation due to pesticide use can be a problem elsewhere. However, it has a large population of some 20 million individuals across its extensive range, and is assessed as Least Concern.

Above left: Tree-holes excavated originally by woodpeckers make ideal nest-sites.

Above right: Like other pygmy owls, this species is a formidable predator of birds. This one has caught a tyrant flycatcher.

This owl hunts mainly small rodents in summer but takes a more diverse diet in winter.

same length and a sweet tone, lasting anywhere between two and 14 seconds. A more slowly paced version is also sometimes heard. The female has a similar song, but her version is shorter and higher pitched. The alarm call is a series of abrupt barking notes. Begging chicks have a drawn-out, shrill and grating screech, and females soliciting food from their mates give a more full-throated version of this call.

Habitat The Northern Hawk Owl breeds in open boreal forests, primarily coniferous but also mixed pines and birches. It uses open ground such as bogs and moorland for hunting, and needs isolated trees with good all-round visibility for hunting perches. Wanderers could turn up in all kinds of lightly wooded habitats.

Behaviour, hunting and diet This owl may be seen active and hunting at any time of the day or night, and will rest in exposed places, such as the tops of solitary dead trees. Its fast flight means it can escape predators more easily and so has less reason to hide in cover than many other owls. It will bob its head and flick its long tail when agitated. It uses exposed perches as watchpoints when hunting, relying more on sight than sound to find prey. It attacks with a steep swooping glide, and may search for prey in flight, hovering before dropping to the ground.

The diet varies through the year, with small mammals of prime importance during the breeding season. These are mainly voles and lemmings – studies on breeding Eurasian birds have found the diet to be about 95 per cent voles by number. In winter, this proportion drops to below 60 per cent, with bird prey making up nearly all of the remainder. It is a powerful hunter and can take prey as large as Sharp-tailed Grouse (in North America) and Willow Grouse (in Eurasia), as well as other, smaller owls. Its hawk-like shape and fast flight make it an effective pursuit hunter of flying birds.

Its ability to hunt birds helps offset the difficulties in finding small mammal prey in the winter – unlike the larger boreal species, such as Great Grey Owl, the Northern Hawk Owl is not adept at locating and catching mammals when they are under snow. It will also take insects and some larger mammals including young hares and weasels.

Breeding This owl does not necessarily maintain its pair bond year after year, nor does it hold onto the same territory through the winter. Adult females in particular are inclined to move away in winter. Males advertise their territories in February and March, singing from assorted perches, and also during a display flight, which will also include wing-clapping. An unpaired female will respond to an advertising male with her own song, and as the bond is formed the two will duet together and indulge in mutual preening.

The male draws his mate's attention to possible

Geographical variation This species is divided into three subspecies. The nominate is found in Europe and most of northern Asia. The subspecies *S. u. tianschanica* replaces it in central Asia and the south-eastern-most parts of the range – this form has more strongly contrasting plumage than the nominate, with the dark areas blacker and the light areas a purer white. In North America is the distinctive *S. u. caparoch*, a darker form with thicker barring on the underside, often an obvious brownish wash on the flanks, a much less pronounced white scapular line and smaller white markings on the upperparts.

Movements and migration As is typical of Arctic owls, the Northern Hawk Owl is a rather nomadic species and is also prone to irruptions when there are severe food shortages. It depends heavily on just a handful of species of small rodents, and these animals' populations tend to rise and fall steeply on a three- or five-year cycle. If they can, adult birds will winter in or at least near their breeding territories, but even in good rodent years juveniles may roam considerable distances. Movements are generally southerly but are unpredictable.

Voice The male's most commonly heard song is a very sustained high-pitched rippling trill, the notes all of the

nest-sites by flying to them and calling. Suitable spots include tree-holes, nestboxes and, less frequently, suitably sized old bird nests, in particular those made by raptors. Holes made by Black Woodpeckers (in Eurasia) or Pileated Woodpeckers (in North America) are large enough to accommodate this species. At about the same time as the nest-site is chosen, courtship feeding and copulation begin, and within a couple of weeks the female is ready to lay her eggs.

Egg-laying usually begins in April. The clutch size is variable, with more laid in good vole years, and northern birds laying more than those further south. The maximum recorded is 13, but between four and eight is more usual. They are laid at intervals of one or two days, but incubation begins as soon as the first egg is laid, so there is a marked difference in chick age. The incubation period is between 28 and 30 days, and during this period the male brings food for the female while she sits, passing it to her at the nest entrance but not entering the cavity himself.

The chicks grow rapidly, kept constantly supplied with food by the busy male, who caches spare prey for times when his hunting is not as successful. This rapid development is a necessity given the short Arctic summer – the earlier they fledge, the longer the chicks will have to work on their hunting skills while prey is abundant. The female remains with the chicks until they are about 13–18 days old, and after that she adds her efforts to hunting. The chicks leave the nest any time between 21 and 32 days, the timing depending on how easily they can climb out and into surrounding branches. They are able to fly when about 35 days old, and need another three weeks or so to become independent.

Status and conservation The Northern Hawk Owl's close association with vole numbers means that its population is always subject to considerable fluctuations. It has the capacity to fully exploit the 'boom' years and produce large broods, helping to offset heavy losses in poor years. It has a very extensive range but is rather thinly distributed, with a total population estimated at 130,000 birds. It is apparently increasing in North America, but appears to be declining in Europe – threats include reduced prey numbers due to pesticide use and loss of suitable nest-sites after forest clearance. Providing nestboxes can improve its fortunes on a local level. It is currently assessed as Least Concern.

Chicks climb into the trees as early as three weeks old.

Asian Barred Owlet
Taenioglaux cuculoides

Right: Some subspecies, including *whitleyi* of China, have streaked rather than barred lower bellies.

The distinctive barred pattern extends to the bird's entire plumage.

Size 22–25cm

Range This owl occurs in south-east Asia, including Taiwan, extending narrowly west through north India and as far as Afghanistan.

Evolution and relationships The genus *Taenioglaux* is sometimes combined as part of *Glaucidium*, but DNA evidence supports its separation. In fact the morphologically quite different genus *Surnia* is closer to *Glaucidium* than is *Taenioglaux*. The Asian Barred Owlet's exact affinities have not yet been studied.

Description Larger than the pygmy owls, this stocky small owl has a longish and full tail. Both upperparts and underparts are dark brown with conspicuous fine white barring. Concentric circles define the facial disk, the chin is white and there are narrow white eyebrows. It has yellow eyes, a pale horn-coloured bill and greyish claws, with feathered tarsi and bare toes. The barred wings are striking in flight. Juveniles have less prominent barring.

Geographical variation There are four or more subspecies. The nominate occurs in mountainous parts in the west of the range. The more reddish *T. c. rufescens* is found in the south-east of the range and has streaks rather than barring on the belly. The form found on Hainan, *T. c. persimile*, is more rufous still, while *T. c. bruegeli* in the north-eastern part of the range is paler and smaller, and also streaked below rather than barred.

Movements and migration It is thought to be sedentary and resident, though mountain populations may move down the slopes in winter.

Voice The male's song is a string of variably spaced sweet notes, which gradually accelerate and become harsher. The usual call is a high pleasant single hoot.

Habitat It breeds in lightly wooded areas (including parks and gardens) in hilly and mountainous areas up to 2,700m.

Behaviour, hunting and diet It is often seen sitting alertly in full view in broad daylight, though it is mobbed by small birds when they find it. It hunts a variety of insects, small mammals and birds, the latter sometimes taken in flight but normally caught on the ground.

Breeding More study is required. In most areas it begins nesting in April. It uses old woodpecker nests or sometimes new ones, driving out or killing the current occupants. The usual clutch size is four.

Status and conservation It is quite common in much of its extensive range, so although its population size is unknown, it is assessed as Least Concern.

Elf Owl
Microthene whitneyi

The Elf Owl is the world's smallest owl species.

Size 13–14cm

Range This owl occurs in the extreme south of the US, reaching into Arizona and the far south-east of California, and in north and central Mexico. There is a separate population in southern Baja California. There may still be a small number of birds surviving on Socorro Island, off west Mexico, but there is general agreement that this population became extinct sometime in the late 20th century.

Evolution and relationships The genus *Microthene* either belongs to or is closely related to the subfamily Surniinae (which holds the pygmy owls, Boreal Owl and Little Owl among others), but it is distinct from the rest of the group and appears to have no particularly close living relatives. The Elf Owl is the sole species in the genus.

Description This is a tiny and delicate owl, the smallest species in the world by weight, although there are a couple of other species that match it by length. It occurs in grey and brown morphs. The upperparts are brown marked with small dark and light blotches on the mantle and shorter coverts, which become dark and light scalloping and then barring on the longer wing feathers and the tail. There is a marked white scapular line. The breast bears soft, broad, brownish streaks, overlaid with fine dark barring, and these thin out to white on the belly. The crown (with no ear-tufts) and nape are finely mottled. It has only a vaguely defined facial disk but quite strong and high-set whitish eyebrows, giving an

intense and rather anxious expression. The eyes are yellow, the bill and claws greyish, and the tarsi are feathered, the toes mostly bare. In flight it looks very petite with short round wings. Juveniles are prominently barred all over.

Geographical variation There are three extant subspecies. The nominate is the most northerly form. The birds on southern Baja California belong to *M. w. sanfordi*, a small and darkish grey form. On Socorro Island it was represented by the more brown and olive-toned form *M. w. graysoni*, considered by some to be a distinct species, but sadly it is now probably extinct. On the mainland in the southern and eastern part of the species' range is *M. w. idonea*, which is greyer and paler than the nominate.

Movements and migration It is a short-range migrant, most birds moving south and towards the coast to winter in southern Mexico, where they can exploit a much larger insect population. In Arizona, the earliest Elf Owl records are in mid-February, and the latest in mid-October. The Baja California population, however, is resident, as were the Socorro Island birds.

Voice The male's territorial and advertising song is an accelerating squeaky chuckle, highest pitched in the middle, lasting about two seconds and comprising about 10 notes that run together at the end of the phrase. It is repeated at intervals of a second or two. He also produces a shorter variant when interacting with a female, for example when encouraging her towards a

Right: Although no larger than a sparrow, this owl occasionally catches other vertebrates.

potential nest-site. The female has her own version of this song, which is shorter and softer than the male's. Other vocalisations include an abrupt high yelp and assorted other high-pitched trills and squeaks. A shrill rolling trill is given by both females and chicks when receiving food, while the chicks give a rasping call when begging for food.

Habitat This is primarily a bird of deserts, and it is particularly associated with the Giant Saguaro cactus *Carnegiea gigantea*, a tree-like, branching species that can grow very tall – more than 13m in some cases – and live for well over 100 years. It is a common species on uplands in north-west Mexico. Areas with a high density of large Giant Saguaro specimens can support dense populations of Elf Owls. However, the Elf Owl also occurs in lightly wooded desert fringes and scrublands where these cacti are absent. In winter, it will use quite open ground without any tall vegetation and will roost among ground vegetation or in bushes.

Behaviour, hunting and diet It is nocturnal, spending the day hiding in holes in cacti or tree trunks or in dense foliage. It becomes active at dusk and is an energetic hunter, searching for prey while flying low over its habitat, hovering in one spot for short spells or running on the ground. It may also sally forth from a low perch, in the style of a flycatcher. Prey is often caught in the air, with a deft flicking swipe of the feet. It will often come to light sources at night, such as camp fires, exploiting the

insects that are attracted to the light.

The Elf Owl is unusual among owls in that it can be heard in flight – it lacks the elaborate suite of sound-dampening feather adaptations of most owls. This reflects its insect diet, as insects are not generally alert to the sound of an approaching predator. It also shows little or no eyeshine, indicating that its eyes have no tapetum, and so its vision in low light may well be poor compared to most owls.

The diet consists mainly of assorted arthropods: insects such as grasshoppers, moths and beetles, and also many spiders and scorpions. Less often it takes vertebrate prey such as rodents and small snakes and lizards, but apparently these are refused when offered to chicks. The exact make-up of the diet varies with the seasons and what is available at different times. For example, in midsummer large beetle species start to predominate as the rains trigger them to emerge from their pupae.

Breeding The classic view of an Elf Owl is a bird peering suspiciously out from a hole in the thick stem of a giant cactus. Often such birds are males, checking out possible nest-sites, which they will later show to their mates. As is usual with migratory species the pair bond may not be maintained over successive years, and indeed may only last three or four months, though the situation may be different among the resident populations.

Among the migrant population, males sing intensively from open perches through moonlit nights as soon as they arrive on the breeding grounds. Once a female comes near, a singing male will move into the mouth of a nest-hole and give his shorter song from there, enticing her to come and inspect the cavity. Over the next few days he will stay close to the female while she moves around and checks all of the nest-sites in the territory, finally choosing the most suitable one. In between this house-hunting, the pair sing and duet, preen one another and copulate frequently. Copulation often takes place immediately after the male has fed the female a prey item.

Most nest-holes used by Elf Owls are cavities made by woodpeckers, usually Gila Woodpeckers, which habitually nest in the cacti, or Northern Flickers, which tend to be tree-nesters. When Elf Owls nest in cacti they enjoy very high fledging success for such a small bird, indicating how safe a home a spiny cactus stem can be.

The average clutch is three eggs, but there may be between one and five, and incubation appears to start when the second egg is laid. The female incubates alone. The incubation period is 21–24 days. The male looks after his mate throughout incubation and this extends to bringing food for the chicks when they hatch, which he passes to the female for her to give to the youngsters. His workload is considerable, and his provisioning visits may be as frequent as once a minute in the peak feeding periods at dusk and dawn. If none of the chicks require food immediately, prey may be left in the nest for later. After a week or two, the female will leave the chicks and add her efforts to the hunting regime. The chicks fledge at about 28 days. They are able to fly straight away, and immediately begin to catch prey for themselves although they receive food from their parents for another week at least.

Status and conservation The Elf Owl appears to be declining in at least parts of its range. Its reliance on woodpecker holes is one of the causes of its vulnerability as it may fare unfavourably in competition with other birds that use similar nest-holes, in particular the introduced European Starling. It is also vulnerable to pesticide use, which can wipe out populations of the larger insects on which it depends. However, with a population of about 190,000 and an extensive range, it is assessed as Least Concern.

Declining and vulnerable to a range of threats, the Elf Owl is still quite widespread for now.

Burrowing Owl
Athene cunicularia

The Burrowing Owl is unusual in that it usually nests in holes underground.

Size 19–26cm

Range This owl has an extensive but patchy distribution across the western half of North and Central America (but also Florida and parts of the Caribbean) and mainly the eastern side of South America but very patchily on this continent. Its northern limits just reach into south-western Canada, while in the far south it reaches into Tierra del Fuego at the southernmost tip of Argentina.

Evolution and relationships The genus *Athene* is part of the subfamily Surniinae, along with the pygmy owls (*Glaucidium*) and Boreal Owl and allies (*Aegolius*). Though now placed in *Athene* as its only New World representative, the Burrowing Owl was formerly classified in a genus of its own, *Speotyto*. However, this separation is not supported by genetic evidence, and now most authorities classify it within *Athene*. Genetic analysis shows that the Burrowing Owl diverged from the Old World *Athene* species some 6 million years ago.

Description A very charming small owl that looks as though it is standing on stilts, the Burrowing Owl's long legs, broad head and upright posture give it a very distinctive outline. The upperpart plumage is mid- to light grey-brown with liberal whitish spots, becoming bars on the primaries, secondaries and tail feathers. The upper breast is brown with pale spots, these becoming larger further down so the pattern is reversed to become dark barring on a whitish ground colour. The facial disk is greyish with a white border, edged at the sides and bottom with blackish. The bill is surrounded with white feathers, and there are heavy pale 'eyebrows' that give the bird a frowning expression. The crown is grey-brown with fine whitish speckles.

The eyes are light yellow, the bill horn-coloured and the claws blackish. The legs are finely feathered to about halfway down the tarsus only. In flight the bird looks quite long-tailed with very broad, round wings. Juveniles have more solid brownish upperparts and a whiter underside.

Geographical variation The nominate subspecies has a large range down the east side of South America. A number of different subspecies from elsewhere in the range have been recognised – up to 21 by some authorities – though the taxonomic validity of some of these is under dispute. The more certain of them include *A. c. floridana* of Florida, which is rather paler and smaller than the nominate, *A. c. hypugaea* of most of North America, intermediate in size between the nominate and *floridana*, the small dark *A. c. troglodytes* of Hispaniola, Beata and Gonave in the Caribbean, a very short-winged and small form *A. c. brachyptera* of northern and central Venezuela and Margarita Island, and *A. c. grallaria*, a quite large and dark form found in the Brazilian interior. In South America in particular, the population is quite fragmented and, as South American populations are generally sedentary, the trend is likely to be towards more genetic distinction between isolated groups.

Right: The obviously long legs are one of this species' striking features.

Little Owl
Athene noctua

Size 21–24cm

Range This owl is found in a broad band across Europe, central Asia and north Africa. Its northern limit in Europe is Denmark and Latvia and it reaches south into Turkey and into the Middle East. Across Asia its range extends south as far as the Himalayas and east to the north-east coast of China. In Africa its main population reaches as far south as northern Mali and Niger, with one or two isolated outposts further south, and more narrowly across northern Libya and Egypt. It has been introduced to England.

Evolution and relationships See Burrowing Owl (page 190) for discussion of the genus *Athene*. The Little Owl is one of four or five Old World *Athene* species and the one with the widest distribution. Of the others, the Lilith Owlet *Athene lilith* of the Arabian Peninsula is its closest relative and is considered by some authorities to be conspecific with the Little Owl.

Southerly Little Owls have paler and sandier plumage than their northern counterparts.

Description A stocky, compact owl with a broad, rather flat-topped head, the Little Owl looks sturdy and quite long-legged. The upperparts are dull grey-brown, the tone varying between subspecies from fairly dark to pale and sandy, with whitish tips to the covert and mantle feathers and whitish bars across the flight and tail feathers, giving a strongly spangled appearance. The underparts are pale, marked with dark streaks that become softer, longer and arrow-shaped on the belly, and coalesce on the upper breast to form a dark upper breast-band. Above this is a whitish collar.

The facial disk is greyish with a white surround and subtle dark edging. There is white feathering around the bill and prominent white brows which create an intense, frowning expression. The crown is grey-brown with whitish speckles. The eyes are yellow, the bill horn-coloured and the claws blackish. The tarsi are feathered white but the toes are bare. In its strongly undulating flight it looks round-winged and top-heavy. Juveniles have more diffuse patterning and a less fierce expression.

Geographical variation The nominate subspecies is found in central Europe and north-west Russia. Several other subspecies are described, though their validity is not agreed upon by all. Further research, especially genetic studies and detailed analysis of vocalisations, is required to properly unravel the species' taxonomy. The form *A. n. vidalli* occurs in western Europe and is darker with more prominent white spotting. In north Africa the subspecies *A. n. glaux* occurs – a slightly smaller pale form that has light cinnamon and grey morphs. In south-eastern Europe the pale *A. n. indigena* occurs, while across into south-western Asia is the pale yellowish-toned *A. n. bactriana*.

Movements and migration Throughout its range this species is sedentary. Young birds disperse an average of 50km from their birthplace.

Voice The male's song is usually written as 'goo-wick' – a long single note that ends with an upward-inflected,

A White Wagtail mobs a resting Little Owl, inviting other birds to join in and help drive it away.

abrupt, almost clicking sound. It is quite high pitched, with a tone that is somewhere between a hoot and yelp. Pairs 'converse' with soft, low calls, and duet together with very cat-like mewing calls. When the bird is agitated its calls become faster and shriller, sometimes reminiscent of the yaps of a small dog. Begging chicks make a harsh hissing call.

Habitat Across its wide range, the Little Owl inhabits a variety of basically open habitats, from farmland to parks, orchards, river valleys, mountain foothills, dunes, clifftops, large rural gardens, semi-desert and woodland edges. It requires at least some trees or other sources of suitable nesting cavities.

Behaviour, hunting and diet Though often seen by day as it may sit inactive in prominent spots, often near its nest-site, this owl does most of its hunting at dusk, and continues to be active into the night. It will also roost by day in more hidden spots within thick tree foliage. Like its relative, the Burrowing Owl, it reacts to an approaching human (or other potential threat) by drawing itself up and bobbing its head, before usually flying away. Its flight, as well as markedly undulating, is often very low, almost skimming the ground.

Its usual hunting mode is the sit-and-wait method, from a perch a metre or so above the ground, such as a fence post, the top of a hedge or wall or a low branch. When prey is spotted it descends in a steep pounce or shallow glide and catches the prey on the ground. It may hawk insects in flight.

The diet is diverse; anything small enough to be caught will feature, although insects make up a large proportion in the summer months especially, and in more southern parts of the range. It will also take earthworms, rodents, young rabbits, frogs, lizards and the occasional bird (including species approaching the owl's own size and weight). The largest prey is generally taken by adults which have well-grown chicks to feed. If prey supplies are good, it will cache extra prey in nooks and crannies for later consumption – these caches may contain dozens of food items.

Breeding As with most other sedentary owls, Little Owls will form lasting pair bonds, both birds defending a rather small territory based around a nest-site. The two birds may be seen sitting together close to their nest at any time of year, though bonding behaviours like mutual preening are most likely to be observed just prior to the breeding season.

Although it preys mainly on invertebrates this owl is quite able to handle mammalian prey.

The pair sticks together all year, with courtship and mating building up from late winter.

The male sings most vigorously in spring, delivering his song from in, or close to, the nesting cavity, and the female may join in, the pair producing a duet (see 'Voice'). Suitable nest cavities include tree-holes, accessible sheltered spaces in derelict or undisturbed farm buildings, and even the burrows of rabbits and other tunnelling mammals. It will readily accept suitably placed nestboxes. Competition for the best nest-sites can be considerable and involve various species, some of which (such as Kestrel and Barn Owl) are larger and therefore capable of ousting a pair of Little Owls. However, some sites are used by successive pairs of Little Owls for decades.

The female lays between three and six eggs at two-day intervals, beginning her incubation on the appearance of the first or second egg. She incubates alone, relying on food deliveries from her mate, and the first egg will hatch after about 22–25 days, the rest following at intervals of a day or two. She remains in the nest with the chicks for a week or two, but begins to join the male in hunting for the family once the chicks are a little more mobile within the nest-hole and are able to keep themselves warm.

Depending on the nest-site's situation, the chicks may be seen outside the nest as early as two weeks of age, though they will generally just sit in the cavity entrance, ready to rush back inside at the first sign of danger. They do not fledge fully until about a month old,

when the wing feathers are well grown enough for flight. They will be fed by their parents for a further month. In a particularly good year, the pair may have a second brood, though a single brood is the norm.

Status and conservation Though it is still a common bird in many areas, the Little Owl is vulnerable to a range of threats, and it has declined in some parts of its range. The main problems it faces include loss of insect prey because of increased use of pesticides and herbicides on farmland and loss of nest-sites as old trees are felled and old buildings patched up. In some parts of its range the latter problem at least is being addressed by providing nestboxes to compensate for losses of other nest-sites. Its total population is placed at between 5 and 15 million individuals, and it is currently assessed as Least Concern.

The Little Owl's status in Britain is an interesting case. Introduced in the 1880s, it established itself throughout most of England, and caused little apparent conflict with native species – perhaps because it already coexisted with them in very similar ecosystems just across the Channel. By the turn of the last century, its population in Britain was – like that of many other farmland birds – in decline. As an introduced species, it will not be targeted by any specific conservation measures, but any initiative aimed at improving biodiversity on farmland should help the species.

Little Owl chicks develop quickly, and by the end of their second week are able to venture to the nest hole entrance.

Boreal Owl
Aegolius funereus

Size 23–26cm

Range This owl, known as Tengmalm's Owl in Europe, has a very extensive range across north-eastern Europe, Russia, China and across Canada and the northern US states, following the conifer belt. There are also small discontinuous populations in suitable habitat in southern and western Europe.

Evolution and relationships See Northern Saw-whet Owl (page 202) for discussion of the genus *Aegolius*.

Description The Boreal Owl is a small species, with short wings and tail and a very broad, square head. The upperpart plumage is grey with large white tips and spots on mantle and wing feathers. The whitish underparts are heavily streaked with grey, the streaks forming arrowhead shapes and converging on the upper breast to give a grey, white-spotted appearance. The pale facial disk is strongly edged with dark grey, and there are bold, prominent black markings on the disk above the eyes, giving the impression of raised eyebrows and a permanently startled expression. The eyes themselves are yellow. The bill is yellowish and the claws are black. The feet and toes are fully feathered. In flight the bird looks short- and broad-winged, heavy-headed and short-tailed. Juveniles are dark brown all over, with pale buff spots on the wing and tail feathers, and whitish feathers around the yellow eyes.

Geographical variation This widespread owl has a number of subspecies across its range. The nominate occurs in Europe and into Russia. In North America the only subspecies found is the larger, darker and more strongly patterned *A. f. richardsoni* (sometimes split as a

separate species). A further three subspecies are found in Asia – in north-eastern Siberia and Kamchatka is the largest form, *A. f. magnus*; in the south-eastern part of the range is the smaller and darker *A. f. caucasicus*; while *A. f. pallens*, intermediate in size between *magnus* and *caucasicus*, is found across most of eastern Asia.

Movements and migration

The more northerly (and, in Eurasia, easterly) a Boreal Owl's breeding grounds, the more likely it is to migrate. Migration is usually short range and may be better classed as nomadic wandering rather than regular migratory habit, as the species' movements are unpredictable and vary according to winter weather and prey availability. In all populations, males are more inclined to remain on their breeding territory through the winter, to avoid having to reclaim it the following spring, while females are more inclined to wander. Young birds of the year are also likely to migrate. Birds on the move in autumn head in a south-easterly direction. In exceptional years, considerable numbers may cross the Baltic Sea, and there have been a few records in Britain.

Voice

Males on territory sing from dusk on still nights from late winter through to early spring to attract a mate. The song is distinctive but shows considerable individual variation, making it straightforward to distinguish different males using the same area of forest. The song is a series of four to nine short hooted notes, each of similar duration and pitch, delivered steadily or rapidly. The owl then pauses before repeating the performance. The song changes when a female approaches to a more extended and stuttering series of notes that become a rapid trill.

Contact calls between the pair include a soft, chick-like cheeping from the female, also used to solicit food

Above: Catching prey in snow is an essential skill for this owl.

Above inset: A very young Boreal Owl chick has closed eyes and a coating of fine white down.

from her mate. The male contacts the female with trilling or soft hooting calls. Vocalisations given in alarm or aggression are loud with a yelping or cracking quality.

Habitat This is a true forest species, and in northern Eurasia is closely associated with natural coniferous woodland of spruce and pine. In North America and southern Eurasia it is equally a bird of deciduous woodland. It will use commercial conifer plantations as well if nestboxes are provided. Northerly populations are found at a range of altitudes but further south it prefers higher ground. It will be excluded from otherwise suitable habitat if there are many Tawny Owls.

Behaviour, hunting and diet As the owl with the most pronounced ear asymmetry of them all, the Boreal Owl is clearly well adapted to a life in the darkness of forests at night, although in North America at least it has been observed hunting by day as well. It spends the daylight hours roosting on a branch against a tree trunk, its dappled grey and white plumage affording good camouflage against the silvery fir-tree bark. If disturbed, it adopts a stretched-up, slender posture to better conceal itself against the trunk, with the crown feathers raised to produce two bumps on top of the head.

It hunts from dusk to before sunrise, perching on a fairly low branch and turning its head slowly from side to side as it listens for movement on the ground. When it detects a moving mammal it drops down onto the prey, sometimes through layers of vegetation. It will move on to a new perch if it hears nothing. The diet is composed almost entirely of small rodents, especially voles – one study on North American Boreal Owls found that 86 per cent of prey taken were voles (92 per cent by biomass). In a bad vole year the owls will hunt small birds, but this does not appear to compensate for the shortage of voles – such years have a dramatic impact on breeding success. When plenty of prey is available food may be cached as a winter larder, sometimes inside a nest-hole, and the owls have been observed thawing out frozen prey by crouching over it as if brooding chicks.

Breeding Male Boreal Owls seek to establish territories that contain at least one good nest-site – in Eurasia this is usually the old nest-hole of a Black Woodpecker, while in North America the nest-hole is likely to be the work of a Northern Flicker or a Pileated Woodpecker. Nestboxes and natural cavities will also be used. Only a small area around the potential nest is defended. Males

begin delivering their advertising song as early as mid-February and no later than mid-March, singing most persistently on still nights when the sound will carry well. Singing reduces after the male has paired up.

An interested female will be 'shown' the nest-hole by the male, who will place a prey item inside as further incentive for her to stay. If after a few inspections she finds the hole suitable she will begin to roost in it by day. The pair meet and copulate each evening and soon the female will lay her first egg and begin to incubate. She lays another every two days until the clutch (of three to seven eggs) is complete. Throughout incubation and the first three weeks of the chicks' lives she remains in the nest, relying on the male to bring food.

Mortality among the younger chicks due to insufficient food supplies is high. The nests are also vulnerable to predation by woodland mustelids (primarily American Martens in North America and Pine Martens in Eurasia), which are agile enough to climb the nesting tree and dextrous enough to extract the contents. A female Boreal Owl responds to scratching sounds on the nest tree trunk by coming immediately to the hole, and attempting to camouflage it by blocking it with her body, but many adult female owls as well as nestlings fall prey to martens. Without any mishaps the chicks will fledge at about 33 days old, when they are capable of weak flight but are competent climbers. The male continues to feed them for up to six more weeks. By the age of three months their post-juvenile moult is complete and their plumage adult-like.

Normally only one brood a year is reared, but successive polyandry has been observed – a female rearing a first brood with one male and a second with another. Polygyny has also been recorded, a male pairing with one female and then another, and provisioning two broods at the same time.

Status and conservation Recording singing males of this species is easy enough in theory, but in reality making an accurate population assessment is extremely difficult, because the breeding range is so vast and includes so much terrain that is difficult to access. The global population is tentatively estimated to be around 2 million individuals, based on known population densities and the approximate area of suitable habitat. There is no obvious general trend of increase or decline, and the species is currently classed as Least Concern. Habitat loss due to deforestation is probably the most significant potential threat.

Scratching sounds on the trunk of its nest tree bring a Boreal Owl quickly to the nest entrance to check for danger.

Left: This owl is usually inactive by day, choosing a concealed spot in which to roost.

Northern Saw-whet Owl
Aegolius acadicus

A Northern Saw-whet Owl with rodent prey. This owl's strength belies its size.

Length 17–19cm

Range This owl is found through the coniferous belt that traverses the northern US, southern Canada (including the Haida Gwai islands, see below) and Alaska and extends south down the western seaboard states. A separate breeding population is found in the highlands of Mexico.

Evolution and relationships The genus *Aegolius* belongs to the subfamily Surniinae, which also includes the genera *Athene* (Little Owl and allies), and *Glaucidium* (the pygmy owls). There are three other *Aegolius* species – Boreal (Tengmalm's) Owl, the Unspotted Saw-whet Owl and the Buff-fronted Owl – all are found in the New World but the Boreal Owl also occurs in Eurasia. DNA studies indicate that the Northern Saw-whet and Boreal Owls shared a common ancestor more than 6 million years ago, while the Northern Saw-whet, Buff-fronted and Unspotted Saw-whet Owls have a more recent common ancestor.

Description This species is small and compact with a broad, squarish head and no ear-tufts. It is smaller and less boldly marked but has proportionately larger wings than the otherwise similar Boreal Owl. The upper-part plumage is grey-brown with white tips to the mantle feathers and coverts, and white spots along the flight feathers and tail. The underside is whitish with heavy but soft and diffuse chestnut-brown streaks. The large but not strongly defined facial disk is dark at the outside, becoming paler towards the centre, and is marked with vague darker streaks radiating outwards. The large round eyes are bright yellowish orange, and the bill and claws are black. The legs and feet are feathered. In flight the owl looks very top-heavy, with short broad wings. Juveniles are darker eyed, with fairly uniform dark brown upperparts and breast, and more orange-toned, unmarked lower breast and belly. Subadults can be aged in the hand by looking at the wing plumage under ultraviolet light, which shows up newer feathers and thus reveals the moult pattern. Some of the juvenile flight feathers may be retained for three or four years.

Left: This owl is active at night, and possesses acute hearing to locate prey.

Right: The Northern Saw-whet is a small owl with a proportionately large head.

Geographical variation There are two subspecies – the nominate is found on mainland North America, while *A. a. brooksi* is restricted to the Haida Gwai islands (formerly known as Queen Charlotte Islands) off the coast of British Columbia. The *brooksi* subspecies ('Queen Charlotte Owl') is considerably darker, more heavily marked, more rufous and darker-eyed than the nominate, and is also resident rather than migratory. A form named '*tacanensis*', recorded from Mexico, is thought to be a hybrid between this species and the Unspotted Saw-whet Owl.

Movements and migration The northern population is partly migratory, appearing in winter in southern and central states where it does not breed. The southern limit in winter just reaches north Mexico, overlapping with the resident southern population, but vagrants may occur outside of this range, reaching central Florida for example. Additionally, those that nest on high ground may move to lower elevations in winter.

Voice The territorial and mate-attracting song is a long sequence of sweet, resonant piping or dripping notes, about two per second, repeated tirelessly through the night in spring (March to May). The usual call is a harsh and rasping screech like a saw being sharpened. A pair interacting together exchange various softer calls.

Larger chicks begging for food give a quite forceful hissing sound. Outside the breeding season this owl is very quiet.

Habitat The northern population of the Northern Saw-whet Owl is found mainly in dense, damp and mature coniferous or mixed woodland, usually at 1,500m or more above sea level. In northern Colorado it will breed to above 3,000m. In all habitats it seems to require a generous understorey, and uses forest edges as it hunts over fields as well as within the woodland. Migrants may use deciduous woodland on lower ground but still favour wet woodlands, composed mainly of willows and aspens, often along riversides. The Mexican population is found in drier and more open habitats.

Behaviour, hunting and diet This owl is a strictly nocturnal species, relying on its camouflage to avoid detection during the day. It prefers to roost in thick and well-sheltered vegetation and often picks a small tree as a roost, which gives it easy access to the lower levels of nearby taller trees for a successful escape flight. If disturbed at its roost, it adopts a sleeked-down conceal-ment posture and partly opens one wing across its body like a shield, as extra camouflage. It mainly uses the sit-and-wait hunting method, moving between several elevated perches, and hunts both within woodland and over adjoining open country. It has very well-developed hearing and probably relies heavily on sound cues to find prey – its relatively low wing-loading also suggests that it is adept at manoeuvring among thick scrub close to the ground.

It mainly takes small mammals, especially woodland mice (which comprise between 40 and 70 per cent of the total diet by number) but also voles, shrews and lemmings. More unusual prey includes small woodland birds caught at their roosts, and occasional insects. Showing power and prowess beyond its size, it has also been noted taking prey heavier than itself, such as Rock Dove and flying squirrel species. The subspecies *brooksi* may visit seashores and take marine crustaceans.

Breeding The male defends a territory with one or more suitable nest-sites and sings from March to attract a mate. On clear and fairly still nights the singing may be almost continuous from dusk to dawn, though it falls silent in strong winds, heavy rain or snow. Pair bonds are not thought to persist from year to year, as is often the case with migratory species, though in the more southerly sedentary populations there is probably more consistency with nest-sites and therefore more chance that the same pair will reform.

The male courts a female with a circular display flight and a perched display of bobbing and 'dancing' along a branch. Sometimes the display ends with him offering her a prey item. He then shows her the nest-hole or holes he has found. These are usually the old nests of woodpeckers, especially flickers but sometimes those of Hairy Woodpeckers. Hole entrances of around 7–9cm are suitable, although larger holes are more vulnerable to predator attacks. The biggest risk to the nest is a raid by a marten, and so the best nest-sites are fairly high up and isolated from nearby branches which would allow a marten to easily jump across.

Eggs are laid between April and July, one every two days. The female begins incubating straight away, so the clutch of between four and seven eggs hatches asynchronously. Younger chicks in larger broods are significantly less likely to survive to fledging age and will be eaten by their siblings if they are starving.

The male provides food while his mate is incubating, and hunts for the family once the eggs have hatched, while the female remains at the nest-site, ready to defend it from climbing mammalian predators. Any sound of scratching on the nest tree brings her immediately up to the nest entrance hole – she may attempt to hide the hole by blocking it with her camouflaged body plumage.

She stops roosting in the nest when the chicks are about 18 days old, and leaves the male to take sole charge of the brood – she may at this time pair with a second male and begin another brood. The chicks are ready to leave the nest at about 33 days old, when they are capable of flight. The male continues to feed them for a further month or so. They look quite different from the adults at fledging age in their mainly chocolate-brown juvenile plumage, but begin to moult at about the time they become independent and are adult-like in appearance by about four or five months old. The young owls are ready to breed in the following spring.

Status and conservation Studying this species is difficult because of its habits. The world population is thought to be 2 million individuals, with an estimated 100,000 to 300,000 in the US. However, there has been a small population decline since the late 20th century. Loss of suitable breeding habitat is the most significant threat, and pesticide use on fields bordering suitable woodland habitat can also be very damaging, killing the owl's rodent prey in large numbers. Measures that could help the species, besides minimising habitat loss and pesticide use, include providing nestboxes. Fixing a barrier around the tree trunk above and below the nestbox is an effective way of keeping martens from predating the nest. Despite some evidence of decline, the Northern Saw-Whet Owl's large population and extensive range mean that its current conservation status is Least Concern.

Left: Northern Saw-whets will use nestboxes in lieu of tree-holes.

Below: The fine filoplumes around the bill have an important sensory role.

Brown Hawk Owl
Ninox scutulata

Size 27–33cm

Range This owl has a wide distribution across the Indian subcontinent and broadly across south-east Asia, north to south-east Russia and Japan, and east to the Philippines and Sulawesi. It is also found on a number of small islands and island groups in the south Pacific.

Evolution and relationships The genus *Ninox*, distributed in Australasia and south-east Asia, is the sole member of the subfamily Ninoxinae, one of the three subfamilies that make up the 'typical owl' family Strigidae. DNA work shows that *Ninox* is basal to the Surniinae owls, and its members are all of distinctive and somewhat 'un-owlish' appearance. They are not closely related to the Northen Hawk Owl of the northern hemisphere, despite superficial similarities. The Brown Hawk Owl is most closely related to the Rufous Owl *N. rufa* and Powerful Owl *N. strenua* of Australasia and Australia respectively.

Description This is a long-bodied and long-tailed, small-headed owl with a rather forward-projecting face. There is considerable variation between subspecies. The upperparts are rather plain brown, with darker barring on the flight feathers and tail. There is a variably distinct white scapular line. The underparts are marked with broad, broken brown streaks, sometimes merging into solid brown at the top of the breast. The head and crown are dark with subtle darker streaking. The facial disk is barely defined and mostly brown but shows a paler cross centred on the bill base and framing the inner edges of the eyes.

The feathering around the bill is short, leaving much of the bill visible, including the nostrils. The eyes are yellow, large and rather bulbous-looking. The bill and claws are greyish. The tarsi are feathered and there is slight feathering along the tops of the toes. In flight the long tail and projecting head recall an *Accipiter* hawk. Juveniles are similar to adults.

Geographical variation Eleven subspecies are recognised, some of which may represent full species. The nominate occurs on Sumatra and nearby islands. The small Javan subspecies *N. s. javensis* is very dark sooty brown on the upperparts. On the Andaman and Nicobar Islands is the uniform chocolate-brown *N. s. obscura*. In Japan is the large and pale form *N. s. japonica*, which is sometimes split as Northern Boobook *N. japonica*, while the more reddish *N. s. randi* of the Philippines and other nearby islands is also sometimes treated as a full species: Chocolate Boobook *N. randi*.

Movements and migration The northerly populations are known to be migratory, moving south and in some areas joining resident southerly populations.

Voice Most resident populations have a similar song, a pleasant-toned, strongly upward-inflected hoot, repeated every second or so. The subspecies *N. s. japonica* and *N. s. randi* have different songs – phrases of two or three hoots with no inflection with short pauses between the notes and longer pauses between the phrases. Females respond with similar but higher-pitched songs.

Habitat This woodland owl inhabits a range of habitat types from upland closed forest to parks and gardens with only scattered areas of tree cover. It also breeds in wooded river valleys, pine plantations and lowland rainforest. As long as tall mature trees are present it will live closely alongside human habitation.

Behaviour, hunting and diet It is a nocturnal bird but somewhat active at dawn and dusk as well. It chooses very thick foliage for its roosting spots and usually watches for prey from an exposed perch, flying out to catch passing insects in its bill or talons. Prey may also be caught on the ground, following a swooping pounce. Its very large eyes but lack of typical owl-face structure suggest it is more reliant on vision than hearing when searching for prey.

The diet is mainly made up of flying insects and sometimes bats. It will also take rodents, crabs, small birds and other small animals that come its way.

Breeding The pair bond may be long-lasting. Paired birds will duet often, while unpaired males call persistently for hours on end through clear nights, most intensively just before dawn. Depending on latitude the breeding season begins sometime between March and May, when the female selects a nest-site, usually a natural tree-hole or woodpecker nest-hole but sometimes a crevice among rocks or piled logs.

The clutch is between two and five eggs, northern birds producing larger clutches on average. Incubation starts when the second or third egg is laid. The female incubates alone and is fed by the male throughout. The male provides for the chicks as well at first, the female joining in when they are a week or two old. They fledge at 24–27 days.

Status and conservation This adaptable owl is thought to have a stable population, although numbers are not known. Its large range means it is evaluated as Least Concern.

The huge yellow eyes dominate this bird's otherwise rather plainly marked appearance.

Left: Like other *Ninox* owls, the Brown Hawk Owl looks rather un-owl-like with its small head and long tail.

Long-eared Owl
Asio otus

Size 35–40cm

Range Long-eared Owls occur across the whole northern hemisphere in a broad band that takes in virtually all of Europe and the northern tip of Africa, the Azores and Canary Islands, then across north and central Asia to Japan and on to the northern US and southern Canadian states.

Evolution and relationships See Marsh Owl (page 216) for discussion of the genus *Asio*. The Abyssinian Owl *A. abyssinicus* of central Africa and the Madagascar Owl *A. madagascariensis*, endemic to Madagascar, are both close relatives of the Long-eared Owl, showing several similarities, and may have speciated more recently than the separation of Short-eared and Long-eared Owl some 5 million years ago, but the details of Abyssinian and Madagascar Owl phylogeny have not yet been fully studied.

Description This beautiful medium-sized owl can look rotund or very slim and sleek, depending on whether it is alert or relaxed. It is somewhat similar to the Short-eared Owl but while that species looks rather grey and sandy-toned, the Long-eared has warmer brown- and orange-toned plumage. The upperparts are grey-brown with a whitish line along the scapulars, dark centres to the mantle feathers and coverts, and barring across the primaries and secondaries. The underside is pale gingery-brown with strong dark streaks, becoming longer with crossbars further down the belly. This heavy streaking, extending right down to the lower belly, is a helpful feature to separate it from the Short-eared Owl at a distance and in flight.

The facial disk is warm orange-brown with a black-and-white edge, and there is a cross of whitish feathers around the bill and inner edges of the eyes. The crown is streaked grey-brown. The large dark-centred ear-tufts above the eyes are evident when the owl is perched

and alert, but not usually visible in flight. The eyes are orange or rich yellow depending on the subspecies, the bill and claws are blackish, and the tarsi and bases of the toes are feathered whitish. In flight the long wings are barred on upper- and undersides, with a clear orange 'window' at the primary bases on the upperside. Juveniles are paler with a blackish facial disk.

Geographical variation There are four subspecies. The orange-eyed nominate occurs throughout Eurasia and north Africa. In most of North America the subspecies *A. o. wilsonianus* occurs, a smaller and darker form with yellow eyes, replaced by the paler *A. o. tuftsi* in western Canada. The Canary Islands is home to the smallest subspecies, *A. o. canariensis*, which has reddish eyes and is smaller and darker than the nominate.

Movements and migration Long-eared Owls are migratory in the northernmost parts of their range, but further south they are sedentary unless forced to move away from their breeding grounds because of prey shortages. Birds that do migrate are quite nomadic on their wintering grounds.

Voice The male's territorial or mate-attracting song is a series of a few dozen rather brief, soft single hooting notes, given from dusk to midnight and then less frequently up until dawn. It may be delivered from a perch or in flight – the display flight also includes wing-claps. The female may answer the male's song with a rather harsh yelping or bleating note. The chicks are well known for having a particularly noisy and distinctive begging call, a shrill squeal like a hinge in need of oiling.

Habitat This species needs both woodland and open countryside – the former for nesting and roosting, the latter for hunting. It is therefore most likely to seek small copses, young plantations or woodland edges, alongside farmland, grassland, cleared areas within plantations or other open habitat. It occupies all altitudes from sea level up to the tree line.

Through much of its range it is more common in coniferous than deciduous woodland, although that may be down to its being unable to compete with other larger owls – in Ireland, for example, where it is the only tree-dwelling owl species, it uses all kinds of woodland. In winter, communal roosts form in trees and thick scrub, often in damp areas such as thickets alongside rivers or around lakes. In eastern Europe it is noted for forming large roosts in urban parks and cemeteries.

Behaviour, hunting and diet A nocturnal species, the Long-eared Owl spends the daytime roosting inconspicuously against the trunk of a tree or within the thick tangle of a dense bush. If disturbed at its roost, it draws itself up into a tall, slim posture with ear-tufts fully raised. Winter roosts can involve large numbers of birds – in parts of eastern Europe, roosts of more than 100 Long-eareds may gather in a single tree.

Hunting usually begins within an hour after sunset and continues until about an hour before sunrise, though this varies according to day length at different latitudes through the year. In the far north, pairs with chicks to feed will have to hunt in daylight at times. It has a similar hunting style to the Short-eared Owl, using a

The nictitating membranes help clean and protect the bird's eyes.

slow and low-level quartering flight over open ground, using headwinds to stall and hover over prey before dropping quickly down to seize it. Hunting from a perch also occurs, especially when the weather is not conducive to hunting in flight.

The Long-eared Owl shares the Short-eared's preference for voles – they can comprise up to 80 per cent of the diet by items. Other mammal species make up most of the rest, including mice, rats, shrews and occasionally larger animals such as rabbits and squirrels. Bird prey can constitute 10 per cent of the diet but usually less, with species taken being mainly small but occasionally up to the size of grouse. Each individual Long-eared Owl (when sedentary) tends to rely on a very small number of species, although what those species are varies from place to place.

Breeding Depending on whether they winter on territory or not, males start singing as early as mid-autumn or as late as early spring. They also display with a circular flight, at about the height of the treetops, both before and after the female's arrival. Birds of both sexes that have already bred together will usually seek to return to the same territory and thus often have the same partner for two or more years, though their loyalty is more to the territory than the mate. The female selects a nest-site within the territory – this is usually an old nest made by a crow or Magpie, but sometimes that of a larger bird. Occasionally the nest will be in a hollow on the ground, and nestboxes have also been used, as well as artificial stick platforms. The same nest may be used in successive years, especially if the pair were successful there previously.

For the next few days the female makes repeated visits to the nest, and the male brings her prey, after which the pair copulate and mutually preen one another. The female begins to lay her eggs soon after, starting to incubate straight away. The full clutch of four to eight eggs is laid over a week or more, with intervals of one to four days between each. The female incubates and defends the nest while the male hunts for her as well as himself – this continues until the owlets begin to leave the nest at about three weeks old. When defending the nest the female may perform distraction displays or adopt a striking defensive posture with the wings half raised and spread, making her appear much larger.

The young owls leave the nest before they can fly by scrambling through the branches – using both bill and claws, they are able to climb back into a tree, parrot-fashion, if they fall to the ground. They alert their parents to their whereabouts with a penetratingly shrill squeaking call – listening for this is the easiest way to detect Long-eared Owl nesting activity. Their flight feathers are developed enough for flight at about five weeks of age, and they become independent at about two months.

Status and conservation The Long-eared Owl has a very large continuous geographic range. Monitoring is difficult, because of its secretive nature and nocturnal habits, but in some well-monitored areas (such as the British Isles) it is known to be slowly declining, and the same may be true generally across its range. BirdLife International estimates its population to be between 1.5 and 5 million individuals. Because of its large range and no evidence of serious decline its status is currently classed as Least Concern.

Short-eared Owl
Asio flammeus

Size 33–42cm

Range One of the most widespread of all birds, the Short-eared Owl has a continuous breeding distribution in a broad band across central and northern mainland Eurasia and North America. It breeds patchily in the British Isles and Iceland, and on various other major and minor islands. It is also found in the Caribbean, north-western and, discontinuously, southern South America, and there are mainly sedentary populations (of distinct subspecies) on some isolated islands and island groups south of the Equator. Some northern birds move south or south-west in winter.

Evolution and relationships See Marsh Owl (page 216) for a discussion of the taxonomic position of the genus *Asio*. The Short-eared Owl's taxonomy is complex, with a number of distinctive island forms that have little or no contact with the mainland populations and are clearly in the process of speciation. Some of these isolated subspecies may already warrant elevation to full species status.

Description This is a relatively long-winged and small-headed owl. Across its range, it shows some variation in colour and intensity of markings. It gives the impression of a rather pale greyish-sandy bird, with strong darker markings. The upperpart feathers are light yellow-buff with whitish fringes and dark central markings, becoming bands on the flight feathers. The underparts are whitish-buff with black streaks, heaviest at the top of the breast. The rather faintly defined white-edged facial disk is yellowish with darker streaks that coalesce to form black patches around the inside edges of bright yellow eyes. This dark surround, along with the whitish cross of feathers around the outside edge of the eyes, makes the eyes themselves stand out and gives the bird a very intense and fierce expression. The legs are feathered, but the toes only partially so. The bill and claws are blackish. In flight the long wings show an unmarked yellowish patch or 'window' at the primary bases on the uppersides, while the undersides are pale with a blackish carpal crescent and dark tips, and barring on the flight feathers. Juveniles have more blackish coloration on the facial disk.

Geographical variation The nominate subspecies is found across the main northern range on both sides of the Atlantic. The smaller and darker *A. f. bogotensis* is found in northern South America and Trinidad and Tobago, while in southern South America it is replaced by *A. f. suinda*, which is closer to the nominate in appearance but somewhat darker and more richly toned. The remaining four to seven subspecies are mainly confined to islands or island groups, and there is differing opinion among taxonomists over whether some can be considered distinct enough to be treated as full species. The form *A. f. galapagoensis*, of the Galapagos Islands, is often split as Galapagos Short-eared Owl *A. galapagoensis*. Other island subspecies include *A.f. ponapensis*, of the tiny South Pacific island Ponapé, *A. f. sandwichensis* of Hawaii, and *A. f. sanfordi* of the Falklands.

Movements and migration Northern populations migrate south for winter – for example, northern Euro-

Roosting spots are often on the ground, under overhanging vegetation.

Left: Widespread and partly diurnal, the Short-eared Owl is one of the easier owl species to see.

and not too many trees or bushes. The Marsh Owl may use savanna, farmland, marshes, estuaries and swamps, but its name is something of a misnomer as it is not closely or exclusively associated with wet or marshy places. It does, however, avoid the driest and most rocky terrain. It readily breeds close to towns and villages, and will even exploit this environment by hawking insects that are attracted to artificial lights.

Behaviour, hunting and diet Although not strictly nocturnal, the Marsh Owl tends to be inactive throughout most of the day, usually beginning to hunt at dusk. It roosts on the ground, concealed in vegetation. Outside the breeding season, small numbers may converge on productive feeding grounds, and communal roosts of dozens of birds may form. This gives the advantage of increased vigilance against predators, and birds leaving the roost in the morning may follow others to locate good places to hunt.

It hunts from a low, slow and agile quartering flight, hovering over prey before dropping down. It will also sit and wait on a low perch, perhaps a small bush or fence post, and watch and listen for prey. The diet is normally composed mainly of small rodents such as multimammate mice – as much as 80 per cent by number – but the owl will also take insects and small birds, sometimes chasing and catching them in flight. The diet varies considerably in different habitats and at different times of year. The Marsh Owl is among many African birds attracted by the emergence of winged termites. Surplus prey is cached for later use.

Breeding The social organisation of Marsh Owls varies from place to place and time to time, probably as a direct result of food availability. Termite emergences, for example, may attract particularly large gatherings. Pairs are monogamous and territorial but may aggregate into loose colonies in productive areas. Breeding takes place mainly in spring after the rainy season, but nests with eggs have been found in all months of the year except January, showing the species' ability to breed opportunistically. It may also nest in very close proximity to nests of the African Grass Owl, apparently without any problems for either species.

Courtship behaviour between the pair involves aerobatic chases and circling flights, performed by the male alone or the pair together, with much vocalisation. The nest itself is a cursory affair, just a trodden-down patch of ground, accessed by a tunnel or run through the surrounding long grass (although at least one Marsh Owl nest has been found in Morocco that was above ground, in an old crow's nest). On the nesting 'pad' the female lays a clutch of two to six, but normally three, eggs, one every two days. Incubation begins when the first egg is laid, resulting in asynchronous hatching and a brood of differing ages. The male brings food to his mate throughout incubation and the early life of the chicks, responding to the female's begging calls, and piling it in the entrance tunnel if the family is already well fed. The female remains with the chicks, guarding them from potential danger and regularly brooding them until they wander from the nest. Her nest defence strategies deployed against larger predators include a convincing distraction display, where she flies off the nest and then crashes down into the grass, feigning injury. She will also chase off passing raptors.

The owlets grow very quickly through the vulnerable early days of their lives. They begin to open their eyes at about seven days old and are able to walk and run easily by about three weeks. At this time they leave the nest and scatter, each finding itself a hiding place among the nearby vegetation. This scattering reduces the chances that the whole brood will be lost to a predator. The chicks give a wheezing call to their parents when a food delivery is on the way, and dance about to attract attention when a parent is near. By about 30–35 days old their flight feathers are well grown enough for their first flights, though their bodies are still downy – full body plumage is not complete until about 10 weeks of age. By this time they are hunting for themselves. The post-juvenile moult is over by about eight months.

Status and conservation The Marsh Owl has a very extensive range and is a common bird in many areas, and although no formal assessment of population has yet been made, there is no evidence for decline or increase. Monitoring the species is difficult, because of its nomadic habits. Locally, it is vulnerable to events such as bush fires as well as climate-related fluctuations in prey availability, but it has the potential to quickly exploit favourable habitat changes, and can achieve high breeding productivity when conditions are suitable. The small Moroccan population, however, has declined and contracted its range, apparently because of habitat loss and disturbance. Overall, the species' conservation assessment is Least Concern.

Its open habitat means the Marsh Owl usually roosts on the ground, hunkering down to hide from predators.

REFERENCES

Backhouse, F. Ghost Chasers.
http://audubonmagazine.org/birds/birds1001.html

Bernd, H. 1993. *One Man's Owl.*
Princeton University Press, New Jersey.

BirdLife International. www.birdlife.org/datazone

Blakiston's Fish Owl Project.
www.fishowls.com/index.html

Cramp, S. (ed.). 1985. *Handbook of the Birds of
Europe, the Middle East and North Africa – The Birds
of the Western Palearctic: Volume IV. Terns to
Woodpeckers.* Oxford University Press, Oxford.

Dice, L. 1945. Minimum Intensities of Illumination
Under Which Owls Can Find Dead Prey by Sight.
American Naturalist LXXIX (784).

Everett, M. 1977. *A Natural History of Owls.* Hamlyn,
London.

Härmä, O., Kareksela, S., Siitari, H. and Suhonen, J.
2011. Pygmy Owl *Glaucidium passerinum* and the
Usage of Ultraviolet Cues of Prey. *Journal of Avian
Biology* 42: 89–91.

Hecht, S. and Pirenne, M. H. 1940. The Sensibility
of the Nocturnal Long-eared Owl in the Spectrum.
Journal of Cell Biology 23 (6).

Hull, J. M., Keane, J. J., Savage, W. K., Godwin, S.
A., Shafer, J. A., Jepsen, E. P., Gerhardt, R., Stermer,
C. and Ernest, H. B. 2010. Range-wide Genetic
Differentiation Among North American Great Gray
Owls (*Strix nebulosa*) Reveals a Distinct Lineage
Restricted to the Sierra Nevada, California.
Molecular Phylogenetics and Evolution 56 (1):
212–221.

Johnsgard, P. A. 1988. *North American Owls:
Biology and Natural History.* Smithsonian Institution
Press, Washington.

Kapfer, J. M., Gammon, D. E. and Groves, J. D.
2011. Carrion-feeding by Barred Owls (*Strix varia*).
The Wilson Journal of Ornithology 123(3): 646–649.

Knudsen, E. I. and Konishi, M. 1978. A Neural Map
of Auditory Space in the Owl. *Science* 200: 795–797.

Koivula, M., Korpimaki, E. and Viitala, J. 1997. Do
Tengmalm's Owls See Vole Scent Marks Visible in
Ultraviolet Light? *Animal Behaviour* 54 (4): 873–877.

König, C., Weick, F. and Becking, J.-H. 2008. *Owls
of the World.* 2nd Edition. Christopher Helm, London.

Korpimäki, E. 1989. Mating System and Mate Choice
of Tengmalm's Owls *Aegolius funereus. Ibis* 131:
41–50.

Marks, J. S., Dickinson, J. L. and Haydock, J. 2002.
Serial Polyandry and Alloparenting in Long-eared
Owls. *The Condor* 104 (1): 202–204.

Marks, J. S., Doremus, J. H. and Cannings, R. J.
1989. Polygyny in the Northern Saw-whet Owl. *The
Auk* 106 (4): 732–734.

Martin, G. R. 1982. An Owl's Eye: Schematic Optics
and Visual Performance in *Strix aluco. Journal of
Comparative Physiology* 145: 341–349.

Nilsson, I. N. 1984. Prey Weight, Food Overlap, and
Reproductive Output of Potentially Competing
Long-eared and Tawny Owls. *Ornis Scandinavica* 15
(3): 176–182.

Powerful Owl observation.
www.pbase.com/rob_hynson/powerful_owl

Roulin, A., Kölliker, M. and Richner, H. 2000. Barn
Owl (*Tyto alba*) Siblings Vocally Negotiate Resources.
*Proceedings of the Royal Society: Biological
Sciences* 267 (1442): 459–463.

Snow, D. W. and Perrins, C. M. 1997. *The Birds of
the Western Palearctic, concise edition.* Oxford
University Press, Oxford.

Solonen, T. 2011. Impact of Dominant Predators on
Territory Occupancy and Reproduction of
Subdominant Ones Within a Guild of Birds of Prey.
The Open Ornithology Journal 4: 23–29.

Steyn, P. . *A Delight of Owls: African Owls
Observed.* Jacana Media, Johannesburg.

Sunde, P. and Bølstad, M. S. 2004. A Telemetry
Study of the Social Organization of a Tawny Owl
(*Strix aluco*) Population. *Journal of Zoology* 263 (1):
65–76.

PICTURE CREDITS

PHOTO CREDITS

Bloomsbury Publishers would like to thank the following for providing photographs and for permission to produce copyright material. While every effort has been made to trace and acknowledge all copyright holders, we would like to apologise for any errors or omissions and invite readers to inform us so that corrections can be made in any future editions of the book.

Key t=top; l=left; r=right; tl=top left; tcl=top centre left; tc=top centre; tcr=top centre right; tr=top right; cl=centre left; c=centre; cr=centre right; b=bottom; bl=bottom left; bcl=bottom centre left; bc=bottom centre; bcr=bottom centre right; br=bottom right

BP = Birdphoto.fi; AG = Agami; SH = Shutterstock

4 BP Markus Varesvuo; **5**tr BP Markus Varesvuo; **6**tl BP Arto Juvonen; **8** BP Markus Varesvuo; **9**cr Christian Artuso; **10**b BP Jari Peltomäki; **11**t BP Markus Varesvuo; **12**bl BP Arto Juvonen; **13** Aurélien Audevard; **14**tl BP Jari Peltomäki, c Subharghya Das; **15**t BP Tomi Muukkonen; **16** BP Jari Peltomäki; **17**br Niranjan Sant; **18** BP Markus Varesvuo; **19**tr HY Cheng; **20**bl Andrés M. Domínguez; **21**tr BP Jari Peltomäki; **21**b [illustration] Jasmine Parker; **22**tl Ian Fisher; **23**tr BP Arto Juvonen, br BP Jari Peltomäki; **24** BP Jari Peltomäki; **25**tr BP Markus Varesvuo; **27** BP Tomi Muukkonen; **28** Christian Artuso; **29**b BP Arto Juvonen; **30**l BP Arto Juvonen; **31**cr Mathias Schäf; **32**tl BP Arto Juvonen; **33**tr BP Markus Varesvuo, bc Chris van Rijswijk; **34**tl BP Markus Varesvuo, bl BP Arto Juvonen; **35**b AG Marc Guyt; **36**l Deborah Allen; **37**tl Ian Fisher, tr Jim Burns; **38**t BP Markus Varesvuo; **39**b Matti Suopajärvi; **40**b BP Jari Peltomäki; **41**tr BP Markus Varesvuo, br BP Tomi Muukkonen; **42**l BP Markus Varesvuo; **43**cr BP Arto Juvonen; **44**bl BP Arto Juvonen; **45**b AG Wil Leurs; **46**tl BP Jari Peltomäki, b BP Markus Varesvuo; **47**tr BP Markus Varesvuo, bl José Carlos Motta-Junior; **48** BP Jari Peltomäki; **49**tr Ron Hoff, cr BP Tomi Muukkonen; **50**b BP Jari Peltomäki; **51** BP Arto Juvonen; **52**t BP Markus Varesvuo, b BP Markus Varesvuo; **53**br Steve Huggins; **54** BP Jari Peltomäki; **55**cr BP Markus Varesvuo; **56**b Hugh Harrop; **57**t Jim Burns; **58**bl Rebecca Nason; **59**l BP Tomi Muukkonen, tr Andrés M. Domínguez; **60**t AG Mike Danzenbaker; **61**b BP Jari Peltomäki; **62** BP Jari Peltomäki; **63** cr BP Jari Peltomäki; **64**t BP Jari Peltomäki; **65**tr Deborah Allen; **66**bl BP Markus Varesvuo, br Rebecca Nason; **67**t BP Arto Juvonen; **68** BP Arto Juvonen; **69**tl Jim Burns, tr HY Cheng; **70**b Christian Artuso; **71** BP Markus Varesvuo; **72**tl Jim Burns; **73** Jim Burns; **74** BP Markus Varesvuo; **75**cr BP Markus Varesvuo; **76**b BP Jari Peltomäki; **77**tr BP Markus Varesvuo; **78** Christian Artuso; **79**tr José Carlos Motta-Junior; **80**tl Rebecca Nason, tr BP Arto Juvonen; **81**b BP Jari Peltomäki; **82** BP Jari Peltomäki; **83**tr SH Russell Shively, bcl BP Arto Juvonen; **84**tl Cover © Cliff Wright 1998, *Harry Potter and the Chamber of Secrets* by J K Rowling, Bloomsbury Publishing; **85**b BP Markus Varesvuo; **86**bcl BP Arto Juvonen; **87**t Jose AS Reyes; **88** SH LivingCanvas; **89**cr BP Arto Juvonen; **90**bl SH David Evison; **91**tl SH Jason Searle, tr SH fritz16; **92**tl SH AISPIX by Image Source, SH Braam Collins; **93**b SH pr2is; **94**b SH Susan Flashman; **95** SH Mark Bridger; **97**tl BP Markus Varesvuo; **98**cl Chris van Rijswijk; **99** Robin Chittenden; **100**b Andrés M. Domínguez; **101**tl AG Marten van Dijl, tr Rebecca Nason; **102**b Michael R. Anton; **103**bcr Ian Merrill; **104**t Christian Artuso; **105** Ian Merrill; **106**bl HY Cheng; **107**t Tadao Shimba; **108** Christian Fosserat; **109**cr Andrés M. Domínguez; **110**t Eyal Bartov; **111**t Christian Fosserat; **113** BP Tomi Muukkonen; **114** Paul Bannick; **116** Paul Bannick; **117** Jim Burns; **118** Jim Burns; **119**cr Jim Burns; **120**tl Christian Artuso; **121**tl Jim Burns, tr Jim Burns; **122** Christian Artuso; **123**cl Christian Artuso, bcl Christian Artuso; **124**t Christian Artuso; **125**tr Ian Merrill, br Jim Burns; **126** Jim Burns; **127**tr Christian Artuso; **128**cl BP Markus Varesvuo; **128–129** BP Markus Varesvuo; **130**t BP Markus Varesvuo, bl BP Markus Varesvuo; **131**t BP Markus Varesvuo; **132**t Deborah Allen; **133**bl Jim Burns, br Deborah Allen; **134** Jim Burns; **135**tr Deborah Allen; **136** BP Markus Varesvuo; **137**tr BP Arto Juvonen; **138**t BP Arto Juvonen, bl BP Arto Juvonen; **139**tl BP Arto Juvonen, tr BP Arto Juvonen; **140**t AG Daniele Occhiato; **142**b AG Marc Guyt; **143** Ian Merrill; **144**b Paul Bannick; **145**tr AG Ian Fisher; **146**tl Ronald Messemaker; **147** Subharghya Das; **148**t Mark Piazzi; **149**br Jonathan Martinez; **150**bl BP Tomi Muukkonen; **151** BP Markus Varesvuo; **152**l BP Arto Juvonen; **153**t BP Markus Varesvuo, br BP Tomi Muukkonen; **155** Amir Ben Dov; **157** Jim Burns; **158**b Eric VanderWerf; **159**tl Jim Burns, tr Jim Burns; **160**t Mathias Schäf; **161** Chris van Rijswijk; **162**t Chris van Rijswijk; **163**b Christian Artuso; **164**t BP Arto Juvonen; **165** BP Markus Varesvuo; **166**b BP Arto Juvonen; **167**tl BP Markus Varesvuo, tr BP Jari Peltomäki; **168–169** BP Markus Varesvuo; **169**br BP Markus Varesvuo; **170**b BP Jari Peltomäki; **171**t BP Markus Varesvuo; **172**tl Choy Wai Mun, tr Martin Hale, bl Niranjan Sant; **173**tr Ron Hoff, bl Ron Hoff; **174**tl BP Tomi Muukkonen, bl BP Markus Varesvuo; **175** BP Markus Varesvuo; **176**t BP Markus Varesvuo, bl BP Markus Varesvuo; **177**b BP Tomi Muukkonen; **178**b Ronald Messemaker; **179**tl Jim Burns, tr José Carlos Motta-Junior; **180** BP Jari Peltomäki; **181**tl BP Jari Peltomäki, tr BP Jari Peltomäki; **182**t BP Arto Juvonen; **183**tl BP Jari Peltomäki; **184**l Ron Hoff; **185**t Martin Hale; **186**tl AG Mike Danzenbaker; **187** Paul Bannick; **188**t Jim Burns; **189**bl Jim Burns; **190**tl Ian Boustead; **191** José Carlos Motta-Junior; **192**b José Carlos Motta-Junior; **193**t AG Wil Leurs; **194**b BP Markus Varesvuo; **195** Filip Verbelen; **196**b AG Marc Guyt; **197**t AG Marc Guyt, br Rebecca Nason; **198**t BP Markus Varesvuo; **199**t BP Tomi Muukkonen, cl BP Tomi Muukkonen; **200** BP Jari Peltomäki; **201**tr BP Jari Peltomäki; **202**t Matt Bango; **203**tl Christian Artuso, tr Deborah Allen; **205**tc Christian Artuso, c Matt Bango; **206** Tadao Shimba; **207**tr Filip Verbelen; **208**tl Steve Huggins, tr BP Tomi Muukkonen; **209** BP Tomi Muukkonen; **210**t BP Arto Juvonen, bl BP Arto Juvonen; **211**t BP Tomi Muukkonen; **212** BP Tomi Muukkonen; **213**tr BP Markus Varesvuo; **214**tl BP Jari Peltomäki, tr BP Jari Peltomäki, cr BP Jari Peltomäki; **215**t BP Markus Varesvuo; **216**b BP Tomi Muukkonen; **217**b BP Tomi Muukkonen; **218** BP Tomi Muukkonen

INDEX

ACKNOWLEDGEMENTS

I would like to thank Jim Martin at Bloomsbury for asking me to write this book, and Julie Bailey and Jasmine Parker for seeing the project through to completion. The text was expertly copy-edited by Wendy Smith, Nicki Liddiard produced a wonderful design and layout, while Helen Snaith did an excellent job on the index and many thanks to Sara Hulse for proofreading. Also, thank you to Nigel Redman and the rest of the Bloomsbury team for casting a keen eye over the final pages. I am grateful to Markus Varesvuo, Arto Juvonen, Tomi Muukkonen and Jari Peltomäki along with the many other photographers whose stunning work brings the book to life. This work would not have been possible without the ornithologists and researchers whose long hours of fieldwork and analysis have uncovered so many secrets about owl behaviour and biology. Finally, I'd like to thank my friends, family and most of all Rob.